D1544595

An Introduction to
Heritage Breeds

Saving and Raising
Rare-Breed Livestock and Poultry

by **THE LIVESTOCK CONSERVANCY**

D. Phillip Sponenberg, DVM
Jeannette Beranger
Alison Martin

Storey Publishing

This book is dedicated to the extraordinary people who conserve endangered livestock and poultry by raising them on their farms. Without their commitment to these breeds, our work would not be possible. Thank you.

*The mission of Storey Publishing is to serve our customers by
publishing practical information that encourages
personal independence in harmony with the environment.*

Edited by Deborah Burns and Ann Larkin Hansen
Art direction, book design, and infographics by Alethea Morrison
Text production by Liseann Karandisecky

Cover and interior illustrations by © Carolyn Guske
Back cover photograph by © Caro/Alamy
Interior photography by Jeannette Beranger/© The Livestock Conservancy, except for
   © Alex Bramwell/Alamy, 17; © Beth Hall/Alamy, 13 top; © Bob Langrish, 137;
   © Caro/Alamy, 8; © Food and Drink Photos/Alamy, 89; courtesy of the Irish Draft
   Horse Society of North America, 181; courtesy of Jeannette Beranger through the
   MacLaughlin Family, 45; courtesy of Jess Brown, 43; © Lynn Stone, 23 bottom;
   Dr. Phillip Sponenberg/© The Livestock Conservancy, 27 top; © Tanya Charter
   and Greg Shore/McKenzie Creek Ranch, 39; © Tim Hill/Alamy, 13 bottom

Indexed by Samantha Miller

© 2014 by The Livestock Conservancy

Storey Publishing
210 MASS MoCA Way
North Adams, MA 01247
www.storey.com

*Storey Publishing is committed to making
environmentally responsible manufacturing
decisions. This book was printed on paper made
from sustainably harvested fiber.*

Printed in China by Shenzhen Caimei Printing Co., Ltd.
10 9 8 7 6 5 4 3 2 1

LIBRARY OF CONGRESS CATALOGING-IN-PUBLICATION DATA
Sponenberg, D. Phillip (Dan Phillip), 1953-
   An introduction to heritage breeds / by D. Phillip Sponenberg, DVM, Jeannette Beranger,
Alison Martin of The Livestock Conservancy.
      pages cm
   Includes bibliographical references and index.
   ISBN 978-1-61212-125-3 (pbk. : alk. paper)
   ISBN 978-1-61212-130-7 (hardcover : alk. paper)
   ISBN 978-1-61212-462-9 (ebook) 1. Livestock—Conservation. 2. Livestock breeds—
Conservation. 3. Livestock—Breeding. I. Beranger, Jeannette. II. Martin, Alison, 1962-
III. Title.
SF105.S755 2014
338.1'62--dc23
                                 2013046366

*Special thanks to Tractor Supply Company for its support of the Livestock Conservancy's
Heritage Breed Photo-Documentation Project.*

CONTENTS

INTRODUCTION

What Heritage Breeds Are and Why They Matter

T HE PHRASE "HERITAGE BREEDS" evokes mental images tinged with warm nostalgia for the smaller and more diverse farms of centuries past, when the pace of life was slower than in today's rushed world. Yet those pleasant pictures only scratch the surface of the rich history of the thousands of livestock and poultry breeds around the world, each uniquely tailored to fit a specific local environment, type of farming, and purpose. These breeds from the past, developed through careful selection by farmers over hundreds of animal generations, are now called heritage breeds.

The rich genetic legacy embodied in these breeds is a key to the future of sustainable agriculture, but heritage breeds are now critically endangered. Fortunately, farmers throughout the United States are stepping forward to rescue and maintain heritage breeds. In the process they are discovering the joys of raising these remarkable animals and have come to appreciate how well they fit on a small-scale, sustainable farm. This book will show how you, too, can help conserve the past to secure the future.

What Is a Breed?

A *BREED* IS A GROUP OF ANIMALS that share a common link of history, original ancestors (*foundation*), and overall body type, all of which work together to result in a reasonable degree of genetic uniformity.

Highland cattle

Livestock breeds work best when they fit their environment and match well with production goals.

Purebred animals predictably reproduce their breed type, both in appearance and performance, when mated with one another. This predictability of appearance, ability, and function through generations is the essential key to the importance of breeds, because predictability is what allows successive farmers and ranchers to consistently achieve their specific goal in a selected setting.

That setting is the other key principle. A Jersey dairy cow is a good choice for producing rich milk on a dairy farm in a temperate climate, while a Texas Longhorn is a good choice for meat production in a drier region. Your farming goals will be met best when the breed matches the setting.

Many Possible Goals

Goals for raising livestock vary widely and can include almost as many options as there are farmers. This list shows some objectives common to different regions and species, but it is only a sketch of the many opportunities possible.

- Milk production from grain-fed animals
- Milk production from grass-fed animals
- Specialty cheese production
- Prime meat production
- Meat production from range-raised stock
- Fiber production
- Egg production from housed stock
- Egg production from free-range stock
- Production of elite breeding stock — tailored for excellence in achieving one of the other specific goals!

THE ORIGINS OF BREEDS

The fascinating history of heritage breeds goes back at least ten thousand years, to the handful of small areas of the globe where agriculture began and the earliest farmers first domesticated the now-familiar species of farm animals. As human populations grew, farmers spread out into diverse climates and terrains, bringing along their domestic livestock, which provided essential food, power, and fertilizing manure.

The resources of these early farmers were usually limited — even hay was a luxury for many. As a matter of necessity, owners quickly learned to keep and breed only those individual animals that thrived and provided a good return for the farmer's (often minimal) investment of feed, shelter, and labor. Characterized by limited resources, these farming systems produced breeds of animals that are still productive today as well as uniquely adapted to local conditions, a key feature of heritage breeds.

Breeds Grow from Community

Breed development is an intricate story of the strong and specific connections among animals, land, culture, and the needs of the people who use them. Each time a group of farmers migrated to a new environment, they chose breeding animals for their abilities both to survive and to produce. Every new settlement began with a different set of founding animals, and each had its own combination of climate challenges, terrain issues, and human demands. Each combination of environment, original animal genetics, and human selection produced its own final genetic packages: that is, its own heritage breeds.

The result? Over the entire globe, human communities in a wide variety of environments tested, molded, and perfected thousands of breeds of chickens, goats, sheep, cattle, horses, and other traditional farm animals. This long history of partnership between animals and people often goes even deeper: many heritage breeds also reflect the cultural approaches to survival of various ethnic groups, specifically the different ways in which each group adapted to and used its environment.

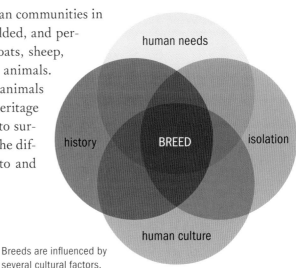

Breeds are influenced by several cultural factors.

Breeds Reflect Their Environment and Their Owners' Needs

Navajo-Churro sheep are survivors. Descended from sheep brought to the New World centuries ago by the Spanish conquistadors, the breed was founded on those few individual sheep that were able to thrive in the harsh desert environment where few other sheep would even survive. Through the generations, the sheep's Navajo and Hispanic herders selected the hardiest, most productive animals for breeding to create today's resilient and productive Navajo-Churro breed. In return, the sheep have faithfully served their owners for several centuries by producing wool and meat. They also came to occupy an important place in Navajo culture and unique roles in Navajo ceremonies.

The distinctive wool has a long outer layer of coarse strong fibers and a softer, short inner layer of very fine fibers. Local Navajo and Hispanic spinners and weavers use it to produce intricate and beautiful rugs that are significant cultural expressions of the two communities most connected with the sheep's history and use. Demand for this unique wool makes the breed important to the health of the local economy and now assures the breed's survival into the future.

The final success or failure of breed survival often depends on animals living on real farms managed by real farmers, who need an economic return from their animals in order to make a living. Successful farmers make for successful breed conservation.

TERMS TO KNOW

GENETIC DIVERSITY
Variation within a breed. Some diversity is obvious, such as color. Less visible traits include disease resistance, foraging ability, and agreeable behavior.

UNIFORMITY
Lack of variation within a breed. Holsteins are fairly uniform for high levels of fluid milk production, and Texas Longhorns for survival in arid regions. On the positive side, farmers can accurately predict performance. The negative side emerges if goals change or the environment changes, because uniform populations may not have the genetic code for adaptation to new challenges.

VARIABILITY
The tendency of traits to vary in response to environmental and genetic influences. Variability can occur within a single breed, so that some animals look and perform differently from others, or among breeds, so that farmers can choose which option best fits their needs.

Navajo-Churro Sheep

Specially adapted to survive and produce in the arid Southwest, the Navajo-Churro sheep is famous for its unique wool and the delicate flavor of its meat.

Homeland: Arid New Mexico, Arizona, Colorado, Utah

Size: Rams 160–200 pounds; ewes 100–120 pounds

Traditional uses: Fleeces for traditional weaving, meat, Native ceremonies

Other traits: Four-horned animals in some bloodlines

Colors: Fleeces come in a warm palette of white, black, grays, browns, and combinations of these. Some colors, such as the beautiful Navajo Sheep Project blue, are unique to the breed.

Conservation status: Threatened

Why Heritage Breeds Are in Trouble

UNFORTUNATELY FOR THE SURVIVAL OF HERITAGE BREEDS, agriculture has changed more in the past century than in the previous 10,000 years. In both developed and developing countries, diversified farming based on adaptation to local conditions is being replaced by the trend to make agricultural systems the same everywhere on the planet. To succeed, this approach requires confinement of animals, standardized feeds, and only the few modern breeds that produce the most in such systems. Thus has the long, historic trend of agriculture toward development of more and more breeds, each better and better adapted to specific regions and multiple purposes, now been reversed. Heritage breeds, and with them their rich *genetic diversity*, are disappearing rapidly.

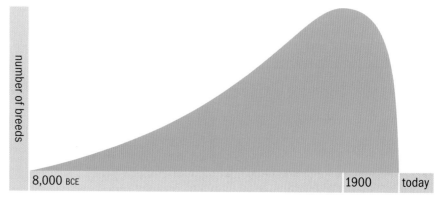

Breed numbers have changed drastically over the past 100 years.

Different Systems Need Different Breeds

Turkeys provide a powerful example of the trajectory of agricultural development over the past century. Originally, all domestic turkeys were *free-range* (not closely confined by fences, pens, or buildings) and kept in outdoor farm flocks where they foraged for food. Industrialization of their production led to humans selecting solely for meat production in a confined situation. Animals were provided all their feed rather than having to seek it out themselves, which meant calories were not burned but instead used to create more meat at a faster rate.

This change in environment allowed for even more selection based strictly on muscle mass, and the final result is the modern Broad Breasted turkey. Because of the massive size of their breasts, toms are no longer able to mount hens properly and so are unable to physically mate. In turn, breeders developed artificial insemination technology in order to propagate the modern turkey. This is a bird produced by industry, unable to rustle up its own lunch and dinner and not even able to reproduce naturally.

The industrial Broad Breasted turkey in a confinement situation is a master at meat production from provided feeds. No heritage breed can match its production in that confined environment. But the industrial bird flunks out in a natural production system where turkeys must naturally reproduce, effectively seek out their own food, and live a long life, thanks to sound muscles and skeleton.

No one type of bird is "best" in an overall sense, but each has its own best setting to provide products unique to its type, background, and ability.

Industrial turkeys are white, and they are confined in houses where all of their feed and water is provided, assuring rapid growth. In contrast, heritage turkeys need to be athletic and able to round up their own resources. They come in a variety of colors, including black, red, bronze, and chocolate.

See pages 14–15 for Breed Snapshots of the Broad Breasted White turkey and heritage Black turkey.

Broad Breasted White Turkey

Adapted for industrial production systems, this breed is noted for its high growth rate and massive meat conformation. The toms' large size prevents them from being able to mount hens properly and mate naturally.

Homeland: United States and Canada

Traditional use: Industrial meat production

Size: Up to 36 pounds for older toms

Color: White

Other traits: Fast growth from provided feeds

Conservation status: Common; commercially available

Black Turkey

Like all turkey breeds of the past — but unlike the Broad Breasted White, opposite — the Black turkey mates naturally and is noted for its free-range foraging ability.

Homeland: United States

Color: Black

Traditional use: Meat

Other traits: Ability to mate naturally

Size: Mature toms up to 33 pounds

Conservation status: Watch

Market Realities

Whenever one breed in a production system is replaced by another breed, demand for the original breed drops, causing its numbers to decrease. Fewer animals mean that fewer and fewer people are familiar with the breed, and the cycle of declining numbers tends to compound until few or no animals of the breed remain. Even more sadly, the decline in use of a breed for production purposes is not always based on fact and measurable differences but rather on a fad for something new. For these reasons many heritage breeds have become extinct, and numerous others are now rare and threatened with that same fate.

THE GLOBAL PICTURE. Global estimates are that 190 livestock breeds were lost forever in just the past 15 years. The Food and Agriculture Organization of the United Nations estimates that 35 percent of livestock and 63 percent of poultry breeds could succumb to the same fate in the next few years.

The trend for breed replacement does have some logic to it: farmers have to make a living, and if your living is based on the productivity of your livestock, then it makes sense to seek out the most productive breeds of livestock for your situation. What is usually overlooked, however, is that the competition between heritage and modern breeds is not on a level playing field. For example, industrial breeds of poultry and swine easily outcompete heritage breeds in high-input, standardized, confined settings, but industrial breeds cannot effectively thrive and produce in the low-input, locally adapted farm systems where heritage breeds thrive. The immediate high-production levels of the industrial system can outshine the long-term security afforded by keeping a wide range of options available for use in future agricultural systems, especially those in which high levels of inputs are no longer realistic.

The allure of the modern breeds is based on productivity under optimal nutrition and management. That underlying high-input system is not always available, or practical, for all farmers, and individuals should be looking closely at whether those breeds can truly be the answer for their production needs. In many cases the answer is "no," and a heritage breed fits better.

Changes in agriculture are certain to occur in the future, just as they have in the past, whether caused by new consumer preferences, climate shift, reduced resources, or things not yet foreseen. Every time a heritage breed is lost, another genetic resource for adapting to future challenges is also lost.

Cultural factors, such as diminishing demand and loss of farmer knowledge, can doom a breed.

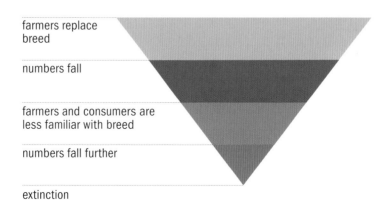

farmers replace breed

numbers fall

farmers and consumers are less familiar with breed

numbers fall further

extinction

Climates Can Shape Breeds

Climate can drive breed survival. Icelandic livestock breeds have gone through multiple periods of near-total collapse as prolonged cold snaps, excessive snow, and even volcanic eruptions have decreased livestock numbers to near-extinction. In each tribulation, only the hardiest survived, and today's Icelandic breeds reflect that adaptation for survival in their genetic makeup.

As the climate becomes more and more erratic in many parts of our country, the ability to produce in a challenging environment becomes key to the survival of individual farms. Having knowledgeable farmers who continue the ongoing, year-by-year process of selecting and breeding those animals that are successfully adapting to local climate change is more important now than it has been for decades.

Desertification has changed environments, forcing farmers to change as well, to drought-resistant cattle and other livestock.

Icelandic Chicken

Icelandic chickens have been selected for 1,000 years for productivity and survival in the cold, dark Arctic winter. Color was not a priority to these breeders and thus varies widely.

Homeland: Iceland

Traditional uses: Eggs, meat

Size: Males, 4.5–5 pounds; females, 3–3.5 pounds

Colors: Too many to list

Other traits: Good foraging ability, good winter egg production

Conservation status: Study

How Small-Scale Sustainable Farmers Can Help Save Heritage Breeds

CONSERVING HERITAGE BREEDS preserves the legacy our ancestors created and handed down as they survived the challenges confronting their lives and their farming systems. Conserving these breeds also preserves our present ability to have animal production work in harmony with the environment. Finally, conserving these breeds preserves our ability to adapt to the future. Because heritage breeds are so well adapted to low-input farming, they offer an ideal opportunity for small-scale sustainable farmers and ranchers to raise resilient, economical livestock and, as a bonus, contribute to the future of sustainable agriculture. Having more farmers make the connection with heritage breeds is the best assurance that all of these breeds will have a future.

Becoming involved with heritage breeds can bring you rewards on several levels. The simple pleasure of working around the animals is extremely satisfying for many owners and is one of their biggest incentives. Knowing that you have helped save a piece of our agricultural heritage for future generations is a huge reward as well. Another bonus is the new people you will meet along the way, and the friendships you will forge as you work together to assure a secure future for your breed. These are all in addition to the economic benefits that many of these breeds offer, essential for farmer survival.

Tracing an American Classic

Dominique chickens are uniquely American, originating here as a productive dual-purpose breed. Once widespread, they have been part of American farm life throughout most of the country for well over a century. The birds still evoke memories for many people, because this was "everybody's grandma's chicken."

The Dominique is a powerful example of how fate can deal real life-or-death challenges to some breeds. The breed was once so common that it was an icon of farm life, but as the nation's egg production moved away from small flocks on every farm and toward large-scale confinement operations using modern Leghorn chickens, the breed dwindled nearly to extinction. Only four flocks remained in 1970. Fortunately for the Dominique and for us, many people have stepped forward with a strong commitment to the breed, so its status is now secure.

See page 20 for a Breed Snapshot of the Dominique chicken.

Dominique Chicken

This historic American chicken has many useful features, most of all its ability to free-range in the typical barnyard or backyard.

Homeland: Eastern United States in 1700s

Traditional uses: Eggs (brown), meat

Size: Roosters, 7 pounds; hens, 5 pounds

Colors: Barred cuckoo color for good camouflage

Other traits: Rose comb resistant to frostbite

Conservation status: Watch

The Livestock Conservancy

THE LIVESTOCK CONSERVANCY, long known as the American Livestock Breeds Conservancy (ALBC), was founded in 1977 by a group of New Englanders concerned about the decline and impending disappearance of several breeds of livestock that were regionally important for both historic and cultural reasons. From a regional beginning, the organization branched out to become a strong national and international voice for the conservation of livestock breeds.

One tool in conserving these genetic treasures is the Conservation Priority List, which sorts the breeds by species and by rarity so that breeders and Livestock Conservancy programs can assure their survival by targeted action.

The thresholds for each category listed below indicate the relative level of both *numerical* and *genetic* threat to the breed.

CRITICAL: Fewer than 200 annual registrations in the United States and an estimated global population of fewer than 2,000 animals.

THREATENED: Fewer than 1,000 annual registrations in the United States and an estimated global population of fewer than 5,000 animals.

WATCH: Fewer than 2,500 annual registrations in the United States and an estimated global population of fewer than 10,000. This category also includes breeds that have special threats to their genetic integrity or a limited geographical distribution, putting the breed at risk from a common disaster such as disease or natural catastrophe.

RECOVERING: These breeds were once in one of the other categories but have fortunately exceeded the criteria for the Watch category. Their numbers still need to be closely monitored to ensure that they do not slip below those criteria.

STUDY: This category is for breeds that are likely to be of genetic interest but for which a full definition of the breed is unavailable, or for which historical or other documentation of the breed is currently inadequate.

If, after reading this book, you decide to join the effort to save heritage breeds, the Livestock Conservancy has extensive additional resources for assisting your efforts. For information and contacts, see Resources.

CHAPTER ONE

Some Background on Breeds

ERITAGE BREEDS ARE IMPORTANT RESOURCES for today's sustainable farmers and for future generations. Just what are heritage breeds and how do they differ from "modern" mainline production breeds? Knowing how these important animals function and how they have become endangered is a great starting point for learning how they can be saved as important pieces of the overall animal production system for the present and the future.

How Breeds Differ

BREED DIFFERENCES CAN BE SUBTLE, but every species also has examples of extreme variation. For example, cattle breeds include:

COLD-HARDY YAKUT that thrive above the Arctic Circle, providing milk, meat, and draft power

WELL-FLESHED AND TENDER ANGUS for beef production in temperate climes

AGGRESSIVE SPANISH DE LIDIA used in the bullring and for beef

DWARF MUTURU of humid tropical West Africa producing milk and beef, and resistant to disease

LONG-LEGGED BORORO that roam the savannas south of the Sahara producing meat and milk for their nomadic owners

ACTIVE AND AGILE THARPARKAR used for draft in India

LONG-EARED GYR used for dairy production in the tropics

BLACK AND WHITE HOLSTEIN unexcelled at milk production in temperate climates

Remarkably different in appearance, these two cattle breeds also have distinctive, valuable traits for farmers. Gyr cattle (top) have seen wide use in the tropics for dairy production. Texas Longhorn cows (bottom) can remain fertile and productive well into their teens.

See page 161 for a Breed Snapshot of the Texas Longhorn.

Each of these breeds has a distinct appearance, adaptation to its environment, and production capabilities suited to the culture of its owners, and each has a role to play in a secure future for global cattle production. Just as important, most would fail to thrive in a different environment due to the mismatch with either the environment, the production goals expected from the animals, or both.

Selecting for Desirable Traits Over the Generations

Each species of domesticated livestock has many breeds. Owners developed these breeds over generations by selecting and breeding animals to emphasize specific *traits* necessary for survival and production in the local environment. Examples of these traits include:

- Strong hooves and good mobility
- The ability to eat and digest fibrous, non-nutritious plants
- The ability to produce a large number of eggs from scrounged feeds

Animals within a breed are relatively uniform and have important differences from those of other breeds.

Breeds organize the total genetic variation of a species into distinct packages, each adapted to a specific purpose in a specific environment. The breeds of a livestock species can be imagined as pieces in a jigsaw puzzle. Each breed fits with the other pieces in the puzzle, working best in its specific spot relative to its neighboring breeds. The entire picture loses something if a piece is missing. Ensuring that we do not lose any more of these breed "pieces" helps to complete the puzzle now and in the future.

TERMS TO KNOW

TRAITS

Inherited characteristics, which can be compared breed to breed and animal to animal. Some of the more important ones include:

- growth rate
- conformation for meat production
- milk yield (amount, or the percentage of fat and protein in the milk)
- fleece length, type, weight, and color
- horns, or lack of horns
- egg size and production
- meat taste
- litter size
- disease resistance
- feed efficiency
- temperament
- mothering ability

SELECTION

Key to successful animal production, selection occurs when breeders allow some animals within a family, herd, or breed to reproduce and some not to do so.

Selection is the most powerful tool available to animal breeders, and by using it they shape the future of their breed. Breeders can select to maintain the traditional type of a breed by favoring typical animals; alternatively, they can radically change the breed by selecting animals with extreme and different type. Selection changes the underlying genetics of the breed by deciding what goes into the next generation.

Exploring the Term "Breed"

Although the word *breed* has several meanings, some based on culture and others on genetics, the definition we will use here is: *A group of animals, developed through isolation and* **selection** *by humans, that share a common link of history, foundation, and type, and that reproduce this same type when mated with one another.*

Each of the terms in this definition has a precise meaning.

HISTORY. The common experience of the animals over generations, including migration, changing use, and management shifts — the practices of function and selection breeders used in the past

FOUNDATION. The specific individual animals that belonged to the original genetic "parents" of the group

TYPE. The characteristic appearance that distinguishes one breed from others

ISOLATION. Lack of mating with animals from outside the original group and its descendants

REPRODUCTION. In this context, the ability to transmit breed type from generation to generation

SELECTION. Breeders' choice of particular animals to carry on desirable traits

HERITABILITY. Consistency of inherited traits in offspring from generation to generation

The most important aspects of the history of each breed are its foundation, isolation, and selection. In all breeds each of these three factors has had a role in the final form of the animals being produced.

In some breeds, one or more aspects may be more important than the others in shaping the final product. For example, Leicester Longwool, Teeswater, and Wensleydale sheep all have a similar foundation, but genetic isolation and the different selection goals of their breeders have shaped them into distinct breeds over time. In contrast, the different foundations of Randall Lineback and Milking Devon cattle caused dissimilarities in these breeds that remain to this day, despite similar selection goals in a similar environment.

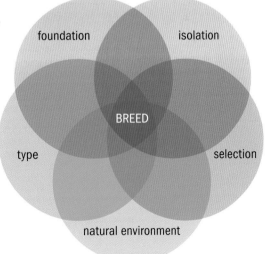

Breeds are shaped by multiple biologic factors.

Important nonbiological factors are also involved with breed definition, such as:

RELATIONSHIP WITH HUMANS FOR SPECIAL CULTURAL USES. An example is the use of four-horned Navajo-Churro rams, raised in the desert, for important roles in Native American "healing way" ceremonies.

GEOGRAPHIC LOCATION WITH ITS OWN DISTINCT ENVIRONMENT. Deserts and humid subtropical environments are likely to be especially challenging for animals. San Clemente goats are adapted for dry hot climates and rocky terrain, and Florida Cracker sheep for hot, humid weather and sandy soils.

Breeds Can Adapt to a Time and a Place

Each heritage breed has its own unique combination of ancestors, environment, and selection history. Different original animals formed the base of each breed, different climates demonstrated which animals had the genetic makeup to thrive, and individual owners from variable cultural perspectives selected animals for specific purposes and characteristics.

For example, Randall cattle, developed in Vermont, originated from a group of cattle from Northern Europe that were imported early in the Colonial period, long before formal breeds had been developed in Europe or anywhere else. Once the cattle were in North America, hardy New Englanders selected breeding animals based almost equally on their milk production, utility as oxen, and ability to survive in that region of long, cold, snowy winters.

In contrast, at about the same time, farmers in the Pineywoods region of the Gulf Coast started with cattle from the Spanish colonies in the region. These cattle came to the Americas before identified breeds were formed in Spain. The local breeders in the region especially needed heat-tolerant oxen to help with logging the local forests. As secondary purposes, these cattle were used for milk and meat production.

The uses of the two breeds were similar, although the cattle themselves are distinct from one another. These differences hail back to the distinct foundation, environment, and selection that are behind each breed.

Breeds Must Not Be Too Genetically Uniform

While a certain level of genetic uniformity is important for predictability, complete genetic uniformity is not the final goal. One of the most important reasons for maintaining some diversity within a breed is that genetic variation allows breeders continually to adapt the breed to changing conditions, whether these are driven by environment or human desires.

A second reason for maintaining some degree of genetic diversity within a breed is that high levels of uniformity come only from *inbreeding*, the mating of closely related animals, which brings with it the real threat of diminished vitality. (See chapters 6, 7, 8, and 9 for more on inbreeding.) Inbreeding results in a group of descendants that are highly uniform due to high levels of genetic similarity. While this sounds good at first, it turns out that some characteristics in animals (especially fertility, disease resistance, and longevity) usually suffer when the animal's genetics become too uniform. These all influence the vitality of the animals.

Similarly distinctive colors can fool you into thinking breeds are closely related, even when they are not. Colorsided cattle pop up in a number of heritage breeds, such as the Randall (top) and the Pineywoods (bottom), even though the breeds are not related at all.

Randall
(Randall Lineback) Cattle

Outstanding contributors to a small farm, Randalls are good foragers, resistant to cold, intelligent and trainable for use as oxen, and provide high-quality rose veal for gourmet cooking.

Homeland: Vermont

Traditional uses: Dairy, meat, draft

Size: Cows, about 1,000 pounds; bulls, 1,800 pounds

Colors: Roan lineback in black (common) or red (uncommon)

Conservation status: Critical

Pineywoods Cattle

Like the Randall Lineback, this historic breed excels as a forager and is adapted to its climate — in this case the heat and humidity of the South. Pineywoods make active, quick oxen and also produce milk and meat.

Homeland: Gulf Coast

Traditional uses: Draft oxen, meat, dairy

Size: Cows, 700 pounds; bulls, 1,200 pounds

Colors: Solid colors and various spotting patterns in black, red, brindle, dun, and dove-gray. Many are speckled and roan.

Other traits: A few are polled, most are horned. Some dwarf types occur in some lines, called "guineas" by their owners and valued for home dairy use.

Conservation status: Threatened

How Genetic Variation Influences Breeding

Not all breeds are created equal! While they all are the result of foundation, isolation, and selection, these play out in different ways for different types of breeds.

Landraces are local breeds that developed from old foundations (200 to 400 years ago in the New World) and achieved their characteristics and adaptations through long-term isolation and selection by their owners. Landraces, at least in their original form, differ from more standardized breeds by having less formal organization of breeders. Landraces achieved their uniformity by the same process as breeds, but in their case the isolation from other animals was usually due to geographic distance or separation imposed by their owners for cultural reasons. The result is a sort of isolation "by default" rather than the more formal isolation "by design" that is typical of more standardized breeds.

For landraces, isolation and *natural* selection are usually the most significant influences on the breed. Unfortunately, the isolation of these animals is no longer consistent, so breeders must now organize and define landraces to preserve them in these days of increased communication and easy transportation.

Standardized breeds are more likely what come to mind when people say "breed." These are groups of animals that share a foundation, isolation, and selection history and have an organized group of breeders who have standardized the look and the breeding practices for the breed. For standardized breeds, the isolation and the *human-driven* half of selection are the most significant parts. Standardized breeds are isolated "by design" so that an Angus cow in England, the United States, or Argentina is part of the same breed genetic pool because matings only occur within the breed.

HOW GENETIC VARIATION AFFECTS BREED TRAITS

GENETIC UNIFORMITY	TRAIT	GENETIC DIVERSITY
High	Predictability	Low
Low	Vitality	High
Low	Response to selection	High
Low	Viability	High
High	Performance uniformity	Low

Diversity within Breeds: Strains, Bloodlines, and Varieties

Subcategories of animals within breeds can be confusing. At what point are these populations part of the same breed and when do they become dissimilar enough to be their own breed? To add to the confusion, the same subcategories sometimes have different names in different breeds and species. But these subgroupings represent much of a breed's diversity, so understanding them and their role in maintaining the vigor of a breed is important.

The most commonly used terms and definitions for breed subgroups are:

BLOODLINE OR STRAIN. A subfamily within a breed, defined as a group of animals within a breed that traces its ancestry back to a specific original group. *Bloodlines* are essentially extended families within a breed. In most cases they derive from a specific family-owned herd that has been relatively closed to outcrosses from other bloodlines in the same breed for some extended length of time. The isolation, as well as the selection by specific owners, usually means that animals from the same strain or bloodline have a specific look or style that sets them apart from the rest of the breed.

In a sense they are *sub-breeds* and have many of the defining aspects of breeds, such as unique foundation, isolation, and selection. They retain the characteristics of the overall breed package, but have the additional stamp of uniqueness that comes from their own group.

Strain has the same meaning as bloodline, and the two terms are used interchangeably.

VARIETY. With a more exact definition than either "bloodline" or "strain," this term indicates the presence of a specific trait or traits possessed by only some of the animals within a breed. *Variety* is most commonly used for poultry breeds to identify birds that differ from one another in comb type or color. For example, Silver Laced Wyandottes and Golden Laced Wyandottes are both varieties of the Wyandotte breed. Similarly, Rose Combed Rhode Island Red and Single Combed Rhode Island Red are varieties of the Rhode Island chicken.

The following three pages compare four distinct bloodlines within the Pineywoods cattle breed.

Hickman Pineywoods Cattle

Homeland: Southern Mississippi, in DeSoto National Forest

Traditional use: Beef production on extensive range

Size: Cows, 700 pounds; bulls, 1,200 pounds

Special adaptations: Rugged, self-sufficient foragers that excel in scrounging food out of the Pineywoods

Colors and other traits: Light, rangy, athletic conformation in a wide variety of colors; horns are usually long and twisted in cows, shorter and stouter in bulls

Population: A few hundred

Conway Pineywoods Cattle

Homeland: Southern Mississippi near Richton

Traditional uses: Draft oxen, beef

Size: Cows, 800 pounds; bulls, 1,400 pounds

Special adaptations: A long history of use as oxen in the logging industry of the Gulf Coast. Teams of up to 5 yokes were used to drag old-growth timber out of the woods

Colors and other traits: Dark red and white in various spotting and roaning patterns; all have moderate-size horns in today's bloodline

Population: A few hundred

Broadus Pineywoods Cattle

Homeland: Southern Mississippi

Traditional use: Beef

Size: Cows, 700 pounds; bulls, 1,200 pounds

Special adaptations: Like other Pineywoods strains, some are dwarf, locally called "guineas." Guinea cattle were highly esteemed for home dairy production because they could meet their maintenance needs more quickly than full-sized cattle

Colors and other traits: Red, brindle, brown, or dun; with or without white spotting. Horns are moderately long

Population: Fewer than 30

Palmer-Dunn Pineywoods Cattle

Homeland: Mississippi-Alabama border

Traditional uses: Beef

Size: Cows, 700 pounds; bulls, 1,200 pounds

Special adaptations: Good foraging ability in the local landscape

Colors and other traits: Red, brindle, dun, many with a white lineback pattern

Many are polled (naturally hornless), which is otherwise rare in Pineywoods cattle

Population: 20

BLOODLINES ADD VARIETY TO BREEDS

Several bloodlines or strains contribute to the uniqueness and variation of Pineywoods cattle. Most of these bloodlines come from specific old families, and many are completely separated from one another by a century of selection and isolation. The result is that Hickman Pineywoods cattle have their own recognizable appearance, which is different from Conway Pineywoods cattle. Each of these, in turn, can be distinguished from Baylis and other Pineywoods cattle strains.

A few lines, such as Palmer-Dunn cattle, continue to have unique variation within a single strain. An example is the existence of both polled (naturally hornless) and horned cattle. Broadus Pineywoods, while all horned, produce the occasional dwarf animal (called *guinea* in the local dialect) as a variation. Some other strains also produce the occasional "guinea" bull or cow. When combined, we see from a larger perspective that each of the bloodlines contributes to the Pineywoods breed picture.

Examples of variation within the Pinewoods breed.

Hickman

Conway

Polled Palmer-Dunn

Broadus

Not All Varieties Are Created Equal

Although the word "variety" has a clearer definition than any of the other subgroup terms, the underlying genetics are muddier. Varieties within poultry breeds are not equally related to one another, and in several breeds the varieties each have a different genetic foundation. Such an occurrence is more possible in poultry than it is in livestock, because the mental image of "breed" for many poultry breeders is a specific final shape, style, and use that is tied more to final appearance than it is to foundation birds. In some cases, individual breeders have used slightly different techniques to come up with a final product that is similar enough for poultry breeders to group the birds together as a single breed. In poultry, some varieties really do function as isolated breeds, rather than as closely related populations separated by one or two genes as seems implied by the word "variety."

On the other end of the spectrum, to complicate matters even more, in some poultry breeds the varieties differ from one another only in the single gene that leads to a color or comb difference. In those breeds the varieties are closely related to one another and do indeed function as variations all within a single genetic breed. Anyone interested in raising poultry breeds with multiple varieties needs to be sure to understand how all the varieties are related in their chosen breed. For poultry varieties, "one size fits all" is definitely not the reality of the situation!

In livestock the picture is clearer, and the formal definition of "variety" is usually not used with mammals. Here, varieties differ less in the underlying genetic package and more in one or two details of appearance, such as the presence or absence of horns, or dwarf versus full size, or different colors. Here are examples of the way breed registries approach such variations:

SPLITTING. In some breeds a difference of variety eventually splits the breed into two differently named breeds. Angus (black) and Red Angus cattle both descend from a single breed but are now separate after long isolation.

COEXISTING. In other breeds, the breed registries allow different colors of the single breed to coexist side by side. Most Oberhasli goats, for example, come in a deep rich bay color (red with black trim), but the breed registry also accepts the solid black goats that are produced occasionally. Randall Lineback cattle are allowed to have a black background color or a red background color. Many heritage livestock breeds allow more variation in color and horn type than is typical of more modern breeds, and this reflects their unique selection history where fancy points had to take a back seat to more practical issue like survival and production.

See page 36 for a Breed Snapshot of the Oberhasli goat.

Oberhasli Goat

This hardy, handsome dairy breed is esteemed for its abundant milk production — up to 4,000 pounds per year.

Homeland: Switzerland

Traditional use: Dairy production

Size: 100–150 pounds

Colors: Red with black trim, or solid black

Conservation status: Recovering

What Distinguishes Heritage Breeds

Most heritage breeds differ from more mainstream production breeds in a few important ways, although the fine details can be confusing. Differences are most dramatic at the two extremes of the spectrum of breeds: landraces and industrial breeds.

Industrial breeds (such as egg-laying chickens, broilers, turkeys, and some dairy cattle) are highly and scientifically selected for maximum production potential in a relatively controlled and benign environment. The result has been exquisitely uniform and productive animals tailored for a specific management system. For industrial breeds only a few animals reproduce, and they reproduce a lot! Successful Holstein bulls, for example, might have tens of thousands of sons and daughters, made possible by the widespread use of artificial insemination. This practice balances the entire breed on a very narrow base of support from these few reproducing animals.

Landraces, in contrast, come to us from a history of local use. Their level of uniformity is relatively low, because each breeder made independent decisions about which animals to reproduce and which animals to cull. The environments for landraces were challenging, and this also imposed selection challenges on the animals — only the survivors could reproduce! Because of isolation from breeder to breeder, the organization of landraces is broader than that of industrial breeds, with each herd or bloodline being its own genetic group.

Landrace animals usually have to juggle adaptation with production, and further have to average their production over a few traits (such as milk, meat, and draft; eggs and meat; or meat and wool) instead of specializing in only one. The result is a balanced animal that is productive "over all" but less likely to be highly productive in any one specialized trait than an industrial animal would be.

Many heritage breeds are closer to the landrace type than to the industrial animal, but they do vary. Some, such as Dutch Belted cattle, Red Wattle hogs, Jersey Buff turkeys, and St. Croix sheep, are fairly specialized in what they produce or do. Others, such as Pineywoods cattle, Spanish goats, and Cotton Patch geese, are generalists and closer to the landrace end of the spectrum. Most heritage breeds fall in between these extremes, with such breeds as Dexter cattle, Romeldale sheep, and Silver Fox rabbits serving as examples of standardized breeds. These lack both the extreme selection for production typical of industrial livestock and also the highly variable situations typical of landraces.

Because nearly all heritage breeds have had a balance of selection for adaptation and production, farmers find them easy to manage and care for while benefiting from their productive potential.

GENETIC UNIFORMITY

HIGH

Industrial breed

Standardized breed

Landrace

LOW

HOW BREED TYPES DIFFER

This chart shows the relative importance of traits in different types of breeds. Heritage breeds are usually more toward the landrace end of the continuum.

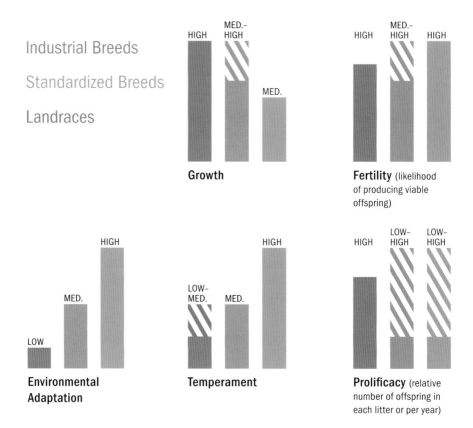

How Environment Shapes Heritage Breeds

Some environments are especially difficult for livestock, and meeting those particular challenges becomes the most demanding and important task for an animal. In other words, the environment can have veto power over all other selection goals.

Humid subtropical environments, for example, present animals with several challenges. Parasite resistance is a top priority for survival. In addition, because forage quality is usually marginal in these environments, animals must be able to select their intake wisely and consume enough to support themselves, their reproduction, and their production. Some examples of breeds that have successfully done this include Florida Cracker cattle, Pineywoods cattle, Guinea Hogs, Gulf Coast sheep, Florida Cracker sheep, and Myotonic goats.

Desert environments can also be challenging, usually in the amount of forage available. The forage present is usually nutritious, just in short supply. Parasites, in contrast, may not be a factor at all. Texas Longhorn cattle, Spanish goats, and Navajo-Churro sheep are all examples of breeds that have met this challenge. These are all thrifty, easy keepers that have good feet and legs to help them navigate the landscape.

Cold places with short growing seasons also impose challenges on both animals and their keepers. Milking Devon cattle, Randall cattle, and Canadienne cattle have all managed to adapt to this situation.

Temperate, lush environments are easier for animals to adjust to, but an adjustment is still necessary. Breeds such as the Dutch Belted and the native Milking Shorthorn have proven valuable for grass-based dairying. Their willingness to get out into the field and graze has been essential in transforming grass into milk without the farmer needing to bring it to them.

In each of these cases, the environment, management, and breed all work together to result in a system that is productive in harmony with the prevailing conditions. By working with, rather than against, the local environment, you can maximize production and minimize the headaches that come from trying to manage the right breed in the wrong place.

Navajo-Churro sheep excel in surviving and producing in the arid Southwest.

See page 11 for a Breed Snapshot of the Navajo-Churro sheep.

How Use Shapes Heritage Breeds

Traditional uses, such as putting cattle to work as oxen, have shaped heritage breeds for centuries. Though some of these functions may no longer be of great importance, their stamp remains on the genetics of the present-day breed because use greatly influences which animals are selected by owners for breeding.

Changes in use, however, are now profoundly impacting both numbers and owner selection criteria in many breeds. For example, draft horses, while still important power sources for some farmers, have declined drastically over the past century with significant effects on animal selection and breed survival. Not only have numbers fallen but owner selection criteria are just as likely these days to favor parade and show use as to reward day-to-day performance in the fields. Draft breeds such as the Percheron and Belgian Draft horse now have taller, rangier bloodlines for show, and shorter, stouter bloodlines for more traditional fieldwork.

American Cream Draft Horses still find a place for their traditional role in field work.

American Cream Draft Horse

The striking color makes this breed stand out, but its power in harness and its smooth action have made it endure.

Homeland: Iowa

Colors: Light to dark cream

Traditional use: Agricultural work

Conservation status: Critical

Size: 1,500–2,000 pounds

How Demand Shapes Heritage Breeds

Changing consumer demand has also drastically changed many breeds by dictating which traits are selected for in the breeding programs of *producers* who need to make a living from their farming or ranching. For example, hogs were historically used as much for lard production as for meat production. Development of hydrogenated oils in the mid-20th century and greater availability of petroleum products "put lard out of a job" in food production, as well as in many other industrial settings. Those trends pressured owners to select for a different sort of hog in the interests of efficient production, one that produced more meat and less (now profitless) fat. This changed the preferred shape and body composition of the ideal modern pig, which in turn has had a definite impact on the underlying genetics of many heritage hog breeds.

The industry-driven alterations in pigs coincided with the advent of total-confinement production systems. This amplified the pressure to select primarily for meat production, and the combined result is a different hog today than a century ago. If the past century has anything to teach us, it's that imagining the hog of the future (or any other animal) is tricky, so saving as many candidate breeds as possible is a wise insurance policy for future generations.

Breeds Can Adapt to Various Uses over Time

The adaptability of heritage breeds to changing environments, consumer preferences, and past, present, and future uses gives them great value in sustainable agriculture.

The story of the Brown farm is an excellent demonstration of this fact. The Ladner/Brown family first settled in the Gulf Coast in 1811 when Carlos Ladner immigrated to the United States. His son, Sebron Ladner, purchased the land that still remains with family members today.

In those early days, Pineywoods cattle were in high demand because they produced hardy and able oxen that hauled huge trees out of the forests of the southern Gulf Coast. The breed powered the region's important lumber industry and the family's livelihood. The family used the cattle to haul timber and sap for turpentine from the local forests, as well as for a savings account and for food.

For Sebron and others like him, cattle were as good as, if not more reliable than, money in the bank. He bought cattle when he had money; he sold cattle when he needed it.

Pineywoods cattle have supported the Ladner-Brown family for generations by providing oxen for work, beef for the table, and a ready source of cash when needed.

Tenth-generation farmer Jess Brown carries on the family tradition of maintaining distinct herds of woods cattle, sheep, and horses, some descended from livestock brought to the New World by Spanish explorers in the 1500s.

Sebron's daughter married Vernon Brown, whose son, Billy Frank, took over the property in 1968. To this day he continues the family tradition and manages a large herd of Pineywoods in partnership with his son Jess.

Although the great logging and turpentine days have passed on the Ladner/Brown land, the Pineywoods cattle remain, helping Jess manage the land by browsing and grazing the brush down around the enduring longleaf pines on the property. This, and the beef production from the cattle, make them doubly productive and beneficial to the family.

The cattle have kept their value by adapting to the changing needs of their owners and the farm economy.

Why Heritage Breeds Are Disappearing

AGRICULTURE UNDERWENT MASSIVE CHANGES in the United States during the past century. The consequent pressure on farmers and ranchers to select animals primarily for their ability to produce has reduced the numbers of nearly all heritage breeds in the past century. Small population size in any breed causes immediate concern among owners because it reduces the likelihood that the breed will survive, and it also causes a second issue that is less easy to recognize.

When a breed has minimal numbers of animals, it provides much less opportunity for producers to select for enhanced production. This can (and often does) result in a slip in the breed's overall productivity, which contributes to a further decrease in numbers as farmers abandon the breed for more productive — and therefore more economically viable — mainstream breeds. The result is a downward spiral of low demand causing decreasing production and resulting in an ever-declining rare breed.

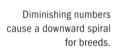

Diminishing numbers cause a downward spiral for breeds.

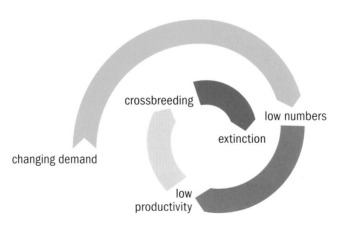

crossbreeding

low numbers

extinction

changing demand

low productivity

How One Breed Went Extinct

The Narragansett Pacer was long considered the premier riding horse of the Americas during the early Colonial period, before major roads were constructed. Known for their smooth gait and endurance, these horses could carry their riders in comfort and cover great distances without tiring. Paul Revere most likely rode a Pacer on his famous ride at the start of the American Revolution.

Developed on the large plantations of coastal Rhode Island, the Narragansett Pacer spawned the growth of a large and robust horse trade in the Northeast, with producers exporting horses throughout the colonies and into the Caribbean. As roads were improved, however, carriage travel grew more fashionable than riding, and demand for the Pacers as riding mounts diminished in the original colonies.

The horses were still in high demand, though, on southern and Caribbean plantations. Plantation owners there, eager for the Pacers, sent agents to Rhode Island to bring back large numbers of horses, including the finest breeding stock. Before the Rhode Islanders realized the loss, a majority of the horses had been taken from the region and by the late 1800s the breed faded into history. The genetics of the Narragansett Pacer breed now remain only in their contribution to several other breeds, including the Standardbred and the American Saddlebred horse, as well as lingering contributions to the few horses remaining in the Caribbean and now crossed with other breeds.

Vintage image of a Narragansett Pacer

How One Breed Was Saved

In 2005 the American rabbit was critically endangered in the United States, until a thoughtful breeding program, initiated by concerned owners, restored the productivity of the breed. Eric and Callene Rapp started with a handful of American rabbits and proceeded to breed them and sort through the litters for the best growth, conformation, and breed type. As a result of these efforts, the popularity of the breed increased as a new market was established for their meat. The Rapps were able to promote the meat as a breed-specific product, coming from a heritage breed.

That link to a high-quality product assured a demand for the breed. By 2012 the American rabbit had moved from being Critically Endangered to Threatened, a significant step toward recovering the breed and its historic importance as a major producer of both meat and pelts. This complete change in direction was the result of careful attention by the Rapps and other breeders to the bottom-line profitability of the breed. The lesson: rabbits that can pay their own way will always be in demand.

How We Can Save Heritage Breeds

THE GOOD NEWS IS THAT BY USING HERITAGE BREEDS in a productive setting, and by reinstituting historical selection procedures, today's farmers can restore heritage breeds to their previous productive role in agriculture. Traditional selection practices were developed by generations of successful producers at a time when these rare breeds were in their heyday. Breeders who promote their animals, and who learn to use techniques with a long track record of success, can readily restore breeds to their traditional production potential. Some breeds have demonstrated surprisingly rapid increases in production levels after only a few generations of selection by their owners. How to do such selection is discussed further in chapter 7.

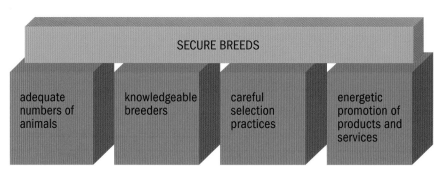

Secure breeds rest on a sturdy support structure.

SECURE BREEDS

adequate numbers of animals

knowledgeable breeders

careful selection practices

energetic promotion of products and services

American Rabbit

This breed boasts everything a rabbit breeder wants: hardiness, docile temperament, good mothering ability, large litter size, and outstanding meat conformation.

Homeland: California

Traditional uses: Meat, fur

Size: Bucks, 9–11 pounds; does, 10–12 pounds

Colors: White, blue; the blue is the darkest of any rabbit breed.

Conservation status: Threatened

CHAPTER TWO

How Heritage Breeds Fit into Landscapes and Farms

ANIMAL AGRICULTURE OF PREVIOUS ERAS was closely and specifically adapted to the world's wonderful array of widely differing environments — from rolling grasslands to steep mountains, humid hills to flat deserts, steaming tropics to Arctic cold, and everything in between. Since few, if any, individual plants or animals can thrive in all environments (how many people do you know who like or tolerate hot and cold weather equally well?), generations of farmers and ranchers developed selection methods to tailor their livestock to fit their local environments. Management styles and production goals also matched each environment. Each combination of environment, selection, management, and goals led to the development of a distinct breed to best exploit what the environment offered and to meet the challenges the environment presented.

In addition to selecting for adaptability, each owner evaluated which animals produced the most as well as the best quality meat, milk, eggs, wool, or labor. These "top of the class" animals — in terms of being both productive and well adapted — were used as breeding stock, each selection working to shape the next generation of animals. This rich heritage of partnership with people, with all

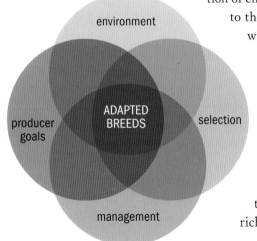

A variety of influences shape adapted breeds.

Spanish Goat, Baylis Bloodline

Breeders developed the Baylis bloodline to strengthen traits that are essential in the Deep South: parasite resistance and efficient use of forage.

Homeland: Mississippi

Traditional use: Local meat production in Mississippi

Size: 60–150 pounds

Colors: Variable

Conservation status: Watch

its complicated interactions through the centuries, shaped each distinctive breed into what it is today.

In this chapter we will discuss the general categories of environment, farming system, uses, and the specific animal traits that are best adapted to each. Breeds function best when their owners manage them as *genetic resources*, meaning that the breed is a collection of genes as much as a collection of animals. This collection of genes is used in a specific environment and under specific management practices, both of which mold that genetic resource and keep the animals relevant to agriculture and viable as a breed.

TERMS TO KNOW

GENETIC RESOURCE

Each breed is made up of individual animals, and each of those animals is made up of different combinations of genes. Some of these genes are consistent across most of the animals of the breed; others are more variable. Both types contribute to the sum total of genetic information in the breed. This sum total is what is available for breeders to work with as they shape the breed for present and future production needs.

The Value of Multipurpose, Adapted Animals

Here are three keys to the past value and future promise of heritage breeds.

1. Their generalist character is highly useful in sustainable farming systems. Heritage breeds have traditionally produced a much broader range of final products than what is now expected from "modern" breeds in large-scale production. Pineywoods cattle, for example, were expected to range widely for forage, work well in a yoke, and produce enough milk to raise a good calf. Holsteins, in contrast, are asked only to produce a lot of milk.

2. The owners selected breeding animals for production as well as for their ability to deal with the climate, to make best use of the available feed, and to adapt to the terrain.

3. Producers selected animals that fit the type of farming they were doing. This selection was critical to how particular breeds developed, and to each breed's success in a particular place and culture.

Landscapes: Climate, Terrain, Soils, and Forages

THE HOME OF ANY BREED IS THE PHYSICAL SPOT it occupies on the globe, combining a host of elements (climate, terrain, soils, and forages) that all contribute to a unique setting. We can group these settings into several general types of environments, each offering challenges and opportunities to the animals residing there. Your goal is to match the breed to your location, so that the animals are well-adapted and productive. A few general ideas can help the process along.

TEMPERATE REGIONS with uniform rainfall are likely to be the most benign environments for most species of livestock. When rain falls, forage grows, and conditions are neither too hot nor too cold, animals are comfortable. Regions of this type promote fast-growing stock that provide top-of-the-line products.

DRY, HOT DESERTS offer animals a host of challenges. Forages in these regions are usually very nutritious, but all those mouthfuls tend to occur far from one another. This puts a premium on tough hooves and good leg conformation for mobility. Desert animals also tend to be on the small side so that they can meet their maintenance nutritional requirements quickly, leaving some reserve left over for production.

HUMID, HOT ENVIRONMENTS are among the most difficult for many animals. Parasites thrive in these conditions, and resistance to them becomes a key component of animal survival. Forages tend to be abundant but low quality, forcing animals to eat a great deal to obtain any nutrition. Feet must be tough in these conditions to resist the challenges of wet soils.

STEEP TERRAIN requires animals to have great mobility and coordination. Good feet and legs are essential for this. The need for mobility in these areas puts limits on how big and broad (and clunky!) animals can be.

COLD ENVIRONMENTS challenge animals with seasonal shortages of forages. While harvested forage in the form of hay or silage can surmount this problem, animals still must consume enough to have the energy to stay warm. They also need to withstand the rigors of cold, and some animals are more up to this challenge than others. Comb type in chickens is a good example, with large single combs more prone than low pea combs to frostbite. Coat type in livestock can also be an important protection against cold.

Gloucestershire Old Spots pigs thrive in temperate settings.

Brahma chickens, with their tight, low combs and small wattles, are very cold resistant.

See page 97 for a Breed Snapshot of the Gloucestershire Old Spots hog and page 52 for the Brahma chicken.

Brahma Chicken

With feathered feet and a small comb resistant to frostbite, the Brahma is well adapted for cold tolerance.

Homeland: United States

Traditional uses: Meat, some egg production

Size: Roosters, up to 14 pounds

Colors: Light, dark, and buff

Conservation status: Watch

Farming Systems

Breeds are closely related to the production settings that formed each one, so understanding these systems is essential in understanding the type of animals best suited to each. Considering the differences between production systems will help you make the right breed choice for your farm or ranch.

There are five broad systems for producing livestock and poultry. Each system values specific traits and pushes animal owners toward selecting for these traits from the available genetic resources. These five are:

- Modern industrial
- Forage-based with extensive range
- Forage-based with high levels of management
- Traditional with concentrate feeding
- Feral

CHARACTERISTICS OF DIFFERENT PRODUCTION SYSTEMS

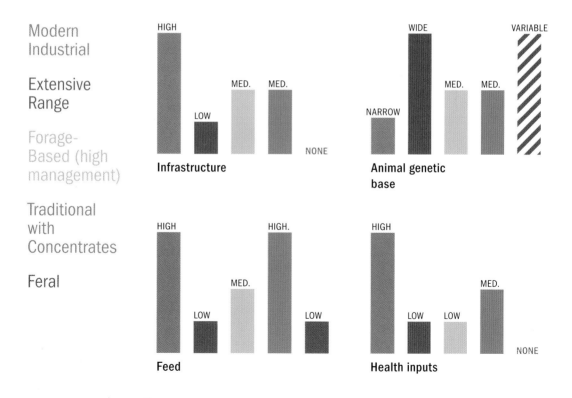

Modern Industrial

Industrial livestock production is the extreme at one end of the range of systems. Producers raise these animals in close confinement, in many cases completely indoors, and provide for all their needs — delivering harvested feeds to their barn, shed, cage, or pen, and removing waste products. This system involves high capital outlay and high operating costs to obtain the highest possible levels of production. Because the animals do not have to seek out their own food or even avoid their own wastes, their bodies focus all their energy on growth.

Influences on breeds in modern industrial systems

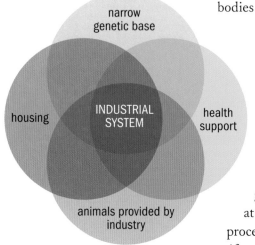

Uniformity and Efficiency

The industrial production model developed fairly recently (mid-1900s) and began with poultry. Vertical integration, in which one corporation owns or controls the entire system, puts the whole production chain from breeder, grower, processor, and final marketing of (in the case of poultry) eggs and meat into that single entity's hands. This system creates uniformity at each step of production: genetics, environment, processing, marketing, and distribution. The result is a uniform product, and the success of this system has led

INDUSTRIAL WHITE LEGHORN

While most chickens could not tolerate the conditions of cage living, breeders developed the White Leghorn to produce efficiently in confinement.

HOMELAND: Now worldwide; originated in Italy
TRADITIONAL USE: Industrial white egg production
SIZE: Hens, 4.5 pounds
COLORS AND OTHER TRAITS: White eggs, white feather color, high egg production (300 or more eggs per year)
CONSERVATION STATUS: Common

many consumers to expect the same uniformity in agricultural products that they expect from mass-produced manufactured goods.

The driving force behind this model of production is efficiency. This system results in the cheap mass production of commodities, which maximizes market share and profits. Although the industrial model has been applied most effectively to chicken (both eggs and meat) and turkey production, hog production is proceeding along this same path, and milk from dairy cows is not far behind.

Differences between Poultry and Beef

Feedlot finishing of beef has also moved toward vertical integration, but so far has been thwarted somewhat at the input end by the thousands of independent producers providing the animals that enter the feedlots. The consolidation of the feedlot step, along with the final slaughterhouse step, of the chain of production has imposed many of the same constraints on beef production as are typical of poultry production. These points of constriction limit the individual cow-calf producer's choice of breeds, because the feedlot and slaughterhouse steps in the system tend to dictate only certain acceptable types. So even for beef cattle, the partial vertical integration of the production system has had broad effects on earlier steps. Still, it lacks the tight industrial control of chicken and turkey genetic resources that characterizes the extreme example of poultry production.

The high reproductive rate of individual hens (nearly an egg a day, and 21 days from egg to chick) also facilitates poultry industrialization, compared to the slower rate of a beef cow (one calf per year). A breeder can thus change a chicken genetic resource by selection (or outright replacement) much more quickly than a beef cattle genetic resource.

Impact on Breed Development

The industrial model has created inexpensive food for millions of people; however, much of the profitability goes to participants other than the farmer. It has also changed the mental image of a productive animal, erasing emphasis on traits of adaptation.

With industrial agriculture, the focus on selection for production above all other traits leaves little room for heritage breeds and almost no room at all for traditional management techniques.

Forage-Based with Extensive Range

Humans developed extensive range production of animals across regions of the globe where crop growing was impossible due to climate (too dry) or topography (too steep). Since these are peripheral areas of modest forage production, animals must get out and walk to find enough to eat. In North America this system is now found mostly where planting crops is simply not an option, such as western ranges or some portions of the Gulf Coast, where land is too steep, rocky, or wet to plow, too dry to grow a crop, or too remote to access easily with farm equipment.

Influences on breeds in extensive range systems

broad genetic base

structural soundess

EXTENSIVE RANGE SYSTEM

longevity

fertility

efficient use of forage

Forage availability in these situations can fluctuate widely with the seasons. In this system, producers can compensate for seasonal deficits by feeding hay or *concentrates* made from the excess of forage in the growing season or brought in from the outside. In the past managers also dealt with these deficits by simply moving the animals, to use mountain pastures in the summer (when the snow had melted) and lowland pastures (and hay harvested from them) in the winter.

ENDANGERED INDUSTRIAL BREEDS?

As a side note, it may surprise you to learn that breeds created for the industrial system can become rare and go extinct just as heritage breeds do. Extinction of industrial breeds usually follows corporate mergers or other restructuring decisions. The loss of these breeds has the same implications for the species and its genetic variation as does the loss of heritage breeds. In the case of industrial breeds, however, little can be done about their loss: the specific environment necessary for their selection cannot be duplicated outside of that industrial setting.

Battery cages for laying hens are deemed undesirable by many people; nonetheless, they are an example of another environmental niche for chickens. These animals are genetically selected to survive and excel in this system. Birds selected for alternative systems would fail in battery cages, and the industrial birds would likewise do poorly outside of the system for which they were created.

TERMS TO KNOW

CONCENTRATES AND CONCENTRATED FEEDS
Feeds based on grains (such as corn) and high-protein legumes (such as soybeans), rather than on grasses and other forages. These range from straight grains (like corn and oats) to pellets made of mixtures of grains, soy, corn gluten, and other sources of protein, supplemented by vitamins and minerals. Concentrates have a relatively high density of nutrients in comparison with grasses and hay.

Extensive range systems, what most people call ranching, transform plants that humans could not otherwise use into animal products we can use: meat, milk, and fiber. Although this is a superb traditional method for making use of otherwise unusable land and plants, it usually occurs in climates and on terrain that are at the extreme of what domestic livestock can use. For this reason, producers in these systems assign a high priority to environmental adaptation of the animals, specifically the traits of long life spans, ruggedness, and productiveness.

Harsh Environments Need Adapted Animals

Arid, expansive ranges are great challenges for livestock producers trying to keep their animals fed, adequately sheltered, and productive. Range also requires careful management because overgrazing and inappropriate grazing can severely damage soils, plant life, and the overall environment. Dry landscapes of the western ranges of the United States are especially vulnerable to damage from overgrazing, and they may take decades to recover from even a few years of poor management. Choosing the right breed can aid greatly in maintaining these environments as well as in having an economically viable ranch.

The right breed also makes a difference when extreme environments become even more extreme. From 2011 to 2013, Texas experienced an intense drought that forced many cattlemen to sell off their livestock because the animals could not support themselves on the remaining forages. Among the few exceptions were the owners of traditional Criollo cattle, such as Sammy Sorsby of the Texas Bar S Cattle Company. His cattle originated in Central Mexico and have been adapted through centuries of natural selection to thrive in the dry and challenging environments of the Southwest and Mexico. Despite the drought, these cattle thrived and even continued to produce offspring on the sparse forage that remained.

Extensive Systems Give Us Unique Landraces

Many North American breeds have been shaped by extensive range conditions, in a variety of environments. Pineywoods cattle, hailing from the humid Gulf Coast, were commonly paired with Gulf Coast sheep to take advantage of the coarse, poor forages produced in marginal wooded areas. Texas Longhorn cattle and many strains of Spanish goats were similarly formed into their present breeds by the rigors of finding enough to eat in the drier areas of the West. In each case the foundation animals had to adapt to the specific local setting, resulting in a final breed that reflected both the foundation stock and the effects of the local environment.

Mexican Criollo cattle thrive on extensive ranges with the fluctuations in forage availability that occur in seasonally dry regions.

Gulf Coast sheep transform the forages of the humid Southeast into meat and wool.

See page 60 for a Breed Snapshot of Gulf Coast sheep.

Mexican Criollo Cattle

Admirably adapted for its environment, the Mexican Criollo can walk long distances, use coarse grasses and other forages, and live a long, fertile life.

Homeland: Northern and Central Mexico

Traditional uses: Beef production, rodeo roping stock

Size: Cows, 600 pounds; bulls, 1,000 pounds.

Colors: Black, red, brindle, spotted, and roan, along with duns and grullo grays.

Conservation status: Study

Gulf Coast Sheep

The traits of this breed make it a boon to Southern farmers: strong feet resistant to damp conditions; parasite resistance; easy birthing; excellent mothering ability; a fertile, long life; and production of both meat and wool.

Homeland: Southern Louisiana, Mississippi, Alabama, and Georgia

Traditional uses: Meat, wool

Size: Ewes, 90–160 pounds; rams, 125–200 pounds

Colors: White, black, or gray

Other traits: Horned with an open spiral, or polled; light fleece of medium wool

Conservation status: Critical

Forage-Based Systems with High Levels of Management

A relatively recent trend in livestock production involves closely managing both animals and forages for grass-based production of grazing livestock and poultry, with minimal or no feeding of concentrates, such as grain. While forage-based animal production goes back for millennia, today's highly managed system, commonly referred to as *rotational grazing* or *management-intensive grazing*, relies on technology, techniques, and philosophies that are recent developments. Some of these philosophies have sprung up in reaction to industrial production's impact on animal welfare and the nutritional quality of that system's final products. Environmental concerns, such as manure production in confinement systems and health concerns from grain-fed meat, have also contributed to the popularity of forage-based systems.

Modern forage-based production maximizes forage growth by subdividing pasture into small paddocks, so each one undergoes brief bouts of intense grazing alternating with long rest periods. This system is made even more productive and environmentally benign if the right breed is matched to the climate and available native forages. Unfortunately, a common temptation for farmers using rotational grazing is to think that any breed will excel in any location. Farmers enthralled with this system have a tendency to be more focused on managing the forage than considering the breed of animal using that forage. Though many breeds within each grazing species do quite well in producing a profit for their owners in a variety of settings, your chances of success improve when you choose a breed that is well-adapted for your region.

Cheap, portable, electric fencing has made the frequent rotation of pastures simple and economical, which in turn allows farmers to maximize forage production, to protect soils from damage, and to accelerate the transformation of grass into meat, milk, and eggs. Many heritage breeds excel in this task, and as a result many of them are in renewed demand as the human health benefits from forage-based animal products are more widely appreciated by consumers.

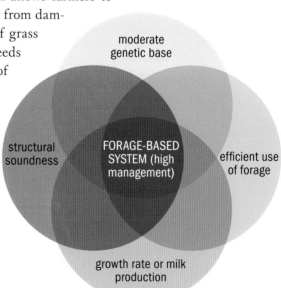

Influences on breeds in forage-based systems with high levels of management

Grass-Based Dairying Needs a Good Foraging Cow

Dutch Belted cattle are an excellent example of a superb grazing dairy breed that was nearly lost and has now been rediscovered. World War II almost spelled the end for the Dutch Belted in their native home of Holland when soldiers in search of food and starving local farmers were forced to eat their dairy cattle. This depleted the population of cattle almost to the brink of extinction. Numbers of the breed also plummeted in the 1900s due to Dutch bull licensing laws. Fortunately, the breed had been imported into the United States by then.

In the United States the breed held its own pretty well up until the dairy buy-out program of the 1980s, when many cattle were sold to slaughter in an effort to reduce cow numbers and raise the price of milk. Interest in the breed grew in the 1990s, both in the United States and in the Netherlands.

Thanks to a number of dedicated breeders such as Kenneth and Winifred Hoffman of Bestyet Dutch Belted Farm in the United States, the Dutch were able to import semen from purebred Dutch Belted bulls to re-establish the cattle back in the Netherlands. A few breeders had held on to their belted cattle, and by coming to the United States for semen from the American branch of the breed, the Dutch breeders were able to give their herds the boost they needed.

The Fine Points of Foraging

Some producers using this type of system also feed minimal amounts of concentrates to maximize productivity of both milk and meat, and to ensure continued animal growth and health during periods of forage unavailability, such as winter or long dry spells. The goal of this system is to maximize productivity from forages, however, so any supplemental feeding is generally kept to a minimum.

For management-intensive grazing to work well, livestock and poultry owners must be attentive to changes in weather and forage growth. A farmer's ability to pay attention to these details usually indicates a talent for managing a breeding program and its associated record keeping, essential to the conservation of heritage breeds. Because rotational grazing is economical, effective, and productive in a wide range of climates, and because most heritage breeds are adapted to be excellent grazers, this system is a logical choice for both sustainable-resource farmers and heritage breeds.

Many grass-based farmers are successfully using crossbreds, hybrids (crosses between two established breeds), or even modern breeds instead of relying on purebred heritage breeds. Though using non-heritage livestock (whether purebred or crossbred) can lead to economic success for the producers, it has the huge drawback of relying on the genetic products of the

Dutch Belted Cattle

The essential traits of good feet and legs get these cattle around to where the grass grows. The cows are fertile, long-lived, and productive into their teens.

Homeland: Netherlands

Traditional uses: Milk production

Size: Cows, 900–1,500 pounds; bulls, up to 2,000 pounds

Color: Black (or rarely red) with a white belt around the barrel

Conservation status: Critical

high-input systems (industrial, high-concentrate feeding) that a rotational grazing system is designed to replace.

The animals destined for production in this system should ideally come from herds selected for performance in this same system. Failing to use heritage breeds in these situations means that these farmers still depend on the modern systems for their genetic material. In a sense they have not "closed the genetic loop" of this production system and are still completely dependent on the availability of less-adapted animals from other systems to fit into their system. By using heritage breeds and selecting them for performance in this system, they are ensuring that the right cattle are available for this system both now and in the future. This closes the loop of the genetic resource (breed) and the production system.

Heritage Breeds Fit Well into Intensive Grazing

Recent expansion of intensive grazing systems has provided renewed interest in breeds such as Devon cattle, which excel at providing high-quality beef from grass. This breed, and its cousin the Milking Devon, have both seen an increase in interest as breeders explore various breeds that fit this system well.

Traditional Systems with Concentrate Feeding

Traditional systems with concentrate feeding are likely to be easily overlooked when considering today's forage-based systems. Concentrates are high-density feeds made mostly from grains, and when these are used to supplement grass, hay, and other forages the result is often rapid growth of high-quality livestock. Major portions of the United States have a long and rich tradition of farmers using livestock to transform forages and grains (usually corn) into high-quality meat, milk, and fiber.

While this system could be considered an extension of the industrial type of management, it is distinct from it and has much older roots. Farmers have been successfully producing animals using this system since the 1800s, and in some regions even before then. Feeding grain to livestock was especially typical in the U.S. Midwest (the "Corn Belt") and for generations has provided much of the income for many small family farms. To make a better profit, a farmer

Influences on breeds in traditional systems with concentrate feeding

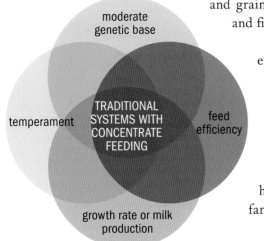

moderate genetic base

temperament

TRADITIONAL SYSTEMS WITH CONCENTRATE FEEDING

feed efficiency

growth rate or milk production

Devon (Beef Devon) Cattle

Favored in many regions for grass-based beef production, Devons are efficient on forages, have good milk production, and produce high-quality beef.

Homeland: Devon, England

Color: Dark ruby red

Traditional use: Beef production

Conservation status: Recovering

Size: Cows, 1,100 pounds; bulls, 1,600–2,000 pounds

could turn corn into either whiskey or meat and, believe it or not, for generations many have chosen meat over whiskey!

This system is adapted to take advantage of rich soils that can support both forages and grains by using a cycle of crop and animal rotations that maintain soil fertility. The harvest was fed to productive breeds of livestock that were selected for rapid growth of tender, flavorful meat. This meat brought a premium price. Concentrates are more nutrient dense than forages, and because they have less bulk it makes them easier to provide to the animals. Feeding concentrates gives animals their basic requirements with less work expended on the animal's part. This frees up extra energy and effort that would have gone into walking and grazing, allowing the energy to go instead into growth and quality of the final product. The products from these animals generally brought higher returns to the farmer than either the forages or the grains could by themselves, and were a convenient way to transform bulky commodities of lower value into those that had greater economic return and were easier to ship to market.

Lush Environments Make for Productive Breeds

Farms in Midwestern North America have rich soils and temperate climates, capable of producing high per-acre yields of grains and forages. These farms are excellent homes for breeds adapted to rapidly transform rich feed into high-quality meat, milk, and eggs. Such livestock have also historically played a key role in maintaining productive landscapes, because they add value to all the crops — corn, small grains, and forages — that farmers traditionally used in their rotational planting systems, and which have proven so effective in conserving the soil's productive potential.

Shorthorn cattle are an excellent fit for rich farmland in temperate climates, and they have a long history of easily transforming good levels of rich feed into milk and beef. In the 1800s the breed began to split into separate lines for beef production and dairy production, and by 1900 that split was formally recognized as different breed associations arose to register the two types separately. The Milking Shorthorn retains that dairy heritage, as do several other heritage breeds of dairy cattle, such as the Milking Devon, Dutch Belt, and Ayrshire.

Like the Shorthorn, several breeds originally had a dual-purpose selection for both dairy and beef production, but in recent decades have seen more of a split between the two types. Red Poll and Devon cattle have seen an emphasis on selection for beef production in the last century, fortunately with the Milking Devon still retaining the dairy potential for that breed. The change of dual-purpose breeds into beef breeds (the more usual pathway) is a reminder that selection can, and has, changed breeds.

Milking Shorthorn Cattle

Classified as Critical on the conservation status list, a purebred individual of this ancient breed is a treasure for the small farmer, providing high-quality milk, beef, and draft power.

Homeland: Yorkshire, England

Colors: Red, white, red roan

Traditional uses: Dairy, beef, draft work

Conservation status: Critical

Size: Cows, 1,200–1400 pounds; bulls, 2,000 pounds

Why Longevity Matters

Traditional systems based on a combination of forages and concentrates take advantage of the genetic potential of livestock to grow quickly on a rich diet. In contrast to the industrial system, this system rewards longevity in addition to consistent high production. Longevity is an important trait in this traditional system because the first few years or months of a brood female's life are nonproductive (before she has that first offspring).

A short productive life means this early nonproductive period is a bigger portion of the entire life than it would be for a long-lived productive animal. So, a cow that lives to 8 years old has produced 6 calves, or 75 percent of her life was productive. If a cow lives to 20 years (yes, this does happen!) and produces 18 calves, then 90 percent of her life was productive.

In addition, the more offspring the animal produces, the more chance for selection of only the very top ones for the next generation. This system also requires moderate levels of adaptation to the surrounding environment because few of these systems are characterized by the complete confinement of animals that is found in the industrial model. As a result, the animals need to adapt to temperature extremes, wet weather, and the levels of sunshine that are present in the environment.

Feral Livestock

Though not technically a farming system, feral animals are an important part of the heritage breeds picture. Feral livestock are domesticated animals that have returned to a free-living state. They often have no owners, and so no specific production goals (which, after all, are imposed by humans). The pure focus on survival assures superb adaptation to local conditions. This can translate into useful roles for these animals when they are brought back into a farm production system. In most cases they have good disease resistance, feed efficiency, and mothering ability, all of which can be used to advantage in productive livestock systems.

Unique challenges face feral breeds, including their survival as well as even being categorized as a breed in the first place. Feral livestock qualify as a breed only if the population has been isolated from outside animals for several decades or centuries. Only a few feral populations meet this requirement, and most of them face ongoing risks to the breed's purity from introduced livestock that can erode their genetic uniqueness.

A second challenge is that many of the islands or other isolated regions where these feral animals live are themselves rare and endangered environments. Feral livestock, by definition an introduced species, can pose a real threat to local indigenous plants and other elements of the environment. In these settings, wildlife conservation programs usually work actively to eradicate any feral livestock. In situations where eradications are planned but where the feral livestock are genetically unique, every attempt should be made by humans to ensure that enough of the animals are returned to domestication so as to allow persistence of their gene pool.

Feral breeds that are successfully transitioned back to domestication still face additional obstacles. One major challenge is that the original survival pressures that shaped them are no longer active. For example, an Ossabaw hog in domestication no longer needs to rely on the annual cycle of acorn abundance, because the farmer will now take responsibility for delivering meals. This means that other selection pressures will be applied either naturally by the new environment or consciously by the new owners. The gene pool will consequently change as it adapts to those new pressures. To effectively save a feral breed in its original form outside of its original native home is impossible, or nearly so.

Another challenge is that feral breeds tend to have fairly low production levels, with the result that few people with an interest in production agriculture think these animals have a role to play in their systems. Their disease resistance, mothering ability, and minimal feed requirements, however, can be exploited in some creative settings, either as purebreds or as a feral base to crossbred production in which the feral females are mated to more standardized breeds for an improved final product. In this case, what is important is for breeders to replenish the numbers of the pure feral breed so that it continues to survive and offer its unique strengths to the production system.

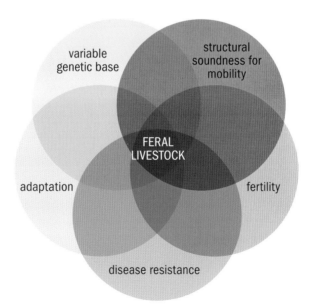

Major factors in shaping feral livestock

Feral Livestock Present Their Own Challenges and Opportunities for Sustainably Minded Farmers

San Clemente goats survived for long years on an isolated island off the coast of California. Slated for removal by the U.S. Navy in 1980, the goats' unique genetic adaptations were in danger of extinction as well. Fortunately, a handful of private breeders have made certain that the genetic line of these goats has continued on the mainland. Even though the environmental and human selection pressures have changed, these breeders continue to work toward maintaining as much of the original genetic package as possible for future generations.

A FERAL BREED ON THE FARM

Maria Castro has been a strong advocate for the San Clemente goat and owns one of the largest herds in Canada. She became involved with the breed when developing her Quinell Lakes Livestock Conservancy farm in Nanaimo, British Columbia. On the farm she raises a diverse group of endangered livestock and poultry breeds that she felt needed attention.

The San Clemente goat was a good candidate for the farm because they are largely self-sufficient and do not need a lot of input from the farm to thrive. To conserve their self-sufficient characteristics, Maria keeps them in large pastures with minimal input of outside feed. The animals are carefully selected for vigor, mothering skills, and proper body type. Any that do not meet the basic criteria expected for this breed within her herd are not used as breeders in following years.

San Clemente Goat

Surviving undisturbed for at least a century on one of California's Channel Islands, this breed thrives with minimal inputs and shows exemplary mothering ability.

Homeland: San Clemente Island, California

Traditional uses: Meat production

Size: Does, 60–80 pounds; bucks, up to 100 pounds

Colors: Most are an attractive combination of black and tan areas, with striped faces and legs. Other colors occur but are rare.

Conservation status: Critical

breed conservation. Farmers may need a lot of off-farm support, but without the farmers the whole enterprise shuts down in a hurry. Choose your species and your breed with care, and you and your chosen animals will be well on the way to success. The right match of farmer and species makes for a fun and rewarding situation for both the animals and the people involved.

Rabbits

RABBITS ARE A NEARLY IDEAL starting species for many people. They are docile, productive, and quiet. Their easy management makes them a good choice for owners of all ages, and young breeders have proven adept at producing top-quality stock. They have few if any zoning restrictions.

The ideal temperature for rabbits is around 50°F (10°C) but they can handle much colder conditions. They are less tolerant of high temperature extremes (90°F [30°C+]) and can become heat stressed unless measures are taken to ensure they don't overheat. Above all else, the area in which the rabbits are housed must be well ventilated.

Size: Small (from 2 pounds to 14 or so; most are 6 to 10 pounds)

Average life span: 5 to 6 years

Handling ease: Easy. Most are docile and size is not an issue.

Noise and odor level: Silent or nearly so. Odor-free if kept clean.

Shelter and space requirements: Small all-wire cages, about 20 × 36 inches × 20 inches (51 × 91 × 51 cm) high, are adequate for many breeds. Some large breeds need a solid floor.

Zoning restrictions: Rarely if ever do these target rabbits.

DAILY FOOD AND WATER REQUIREMENTS. Rabbit feed is widely available in pet and farm stores. Browse, hay, or garden leftovers are welcome but should not be fed if contaminated with pesticides or fertilizer residues. Some weeds are toxic to rabbits. Water is essential, and most housing systems have included this in their planning.

SOCIAL STRUCTURE. Males and females should be housed separately after weaning. They are placed together briefly for mating, then separated again.

REPRODUCTION. Rabbits are induced ovulators, typically responding to mating with ovulation and conception. Most are sexually mature at about 3 to 8 months, and usually at the lower end of this range. Gestation is about a month. Litters vary from four to ten babies (kits) and can be produced every 3 months or so.

PREDATOR CONTROL. Rabbits are vulnerable to a wide range of predators. Cages usually provide adequate protection against most of them.

PRODUCTS. Most conservation priority rabbit breeds are specifically designed to produce meat. Rabbit pelts may be profitable in some niche markets. Pelts of some breeds are in higher demand than others.

PROCESSING AND TRANSPORTATION. Rabbits are relatively easy to cage and ship from place to place when breeding stock is sold to distant customers. Selling large numbers intended for meat processing is impossible in many areas, though, due to a lack of facilities.

POTENTIAL MARKETS. Some small companies throughout the United States purchase rabbit pelts. Local demand for meat rabbits is variable, and can be quite high in some areas.

BREED ASSOCIATIONS AND OTHER RESOURCES. The American Rabbit Breeders Association has tens of thousands of members throughout the United States and Canada. The association organizes shows, offers registration services, and determines breed standards.

OTHER. Manure from rabbits is a gardener's dream and is a highly sought-after fertilizer, especially on organic farms. Further marketing opportunities from manure can come from growing earthworms in the manure and selling the worms for fishing bait.

RABBIT BREEDERS SAVE RARE BREEDS

Eric and Callene Rapp and their Rare Hare Barn rabbitry in Kansas have made significant contributions to the survival of several rare breeds of rabbits including the American, American Chinchilla, Blanc de Hotot, Crème d'Argent, and others. They have also created a highly successful meat and breeding stock business with their lines of rabbits through the careful attention they give to the productivity of their animals and good marketing.

The Rapps select heavily for meat production while making sure the animals meet the breed standards set by the American Rabbit Breeders Association. Other characteristics that they value in their animals include heat tolerance, fertility, and good mothering. The end result of this careful selection process is valuable breeding herds and a prosperous meat business.

Silver Rabbit

The docile Silver is highly valued for both meat and its lush pelt. A hardy, productive breeder, it produces three to six kits per litter.

Homeland: England

Traditional uses: Meat, pelts

Size: 4 to 7 pounds

Colors: Black, brown, or fawn, all with silver hairs scattered in

Conservation status: Threatened

Beveren Rabbit

Valued traits of the Beveren include large litters and rapid growth. The does are good, docile mothers.

Homeland: Belgium

Colors: Blue, black, blue-eyed white

Traditional uses: Meat, pelts

Conservation status: Watch

Size: Does, 9–12 pounds; bucks, 8–11 pounds

Chickens

CHICKENS ARE A GOOD INTRODUCTORY species for many beginners, with many breed options to appeal to a wide variety of tastes. Their prolificacy and short generation interval make them nearly ideal for learning the basics of selection and seeing the results quickly.

Size: Generally small, but variable. Bantams are as small as 1 pound; the largest Brahma roosters are up to 14 pounds.

Average life span: 5 to 7-plus years. (Phil Sponenberg's grandmother had a pet hen that lived 23 years!)

Handling ease: Usually easy to handle, but roosters of some breeds can be aggressive, depending on how they are raised. Some breeds are nervous and flighty, others calm and docile.

Noise and odor level: Roosters crow, and this can be loud in some breeds. Hens chatter and cluck, occasionally at high volume. Odors are minimal if facilities are kept clean, but overwhelming if manure builds up.

Shelter and space requirements: Shelter is needed, especially from rain. A coop about 3 × 5 × 2 feet (1 × 1.5 × 0.6 m) can comfortably hold 5 hens of most breeds.

Zoning restrictions: Regulations excluding chickens are common in many areas, although zoning is changing with the increased popularity of backyard chickens. Restrictions against roosters still remain in many areas.

DAILY FOOD AND WATER REQUIREMENTS. Chicken feeds are widely available at feed stores, and chickens also avidly eat kitchen scraps. Fresh liquid water is a daily requirement and should be available at all times.

SOCIAL STRUCTURE. Chickens are sociable, but the term "pecking order" originated with chickens! Crowding in confined spaces can lead to bird-on-bird aggression. Most breeds require a rooster for every seven to ten hens, and multiple males in a smaller group can worry the hens incessantly. Introducing new adults into a flock can result in turmoil.

REPRODUCTION. Many breeds lay nearly an egg a day depending on nutrition and light levels. Pullets (young hens) usually start laying eggs at about 5 to 6 months old. Incubation is about 21 days.

PREDATOR CONTROL. Essential. Tight housing can eliminate most predation — and a good guard dog will also do the trick.

PRODUCTS. Eggs, meat (fryers from younger stock, broilers from larger stock), feathers for jewelry or fly-tying, and manure for gardeners.

PROCESSING AND TRANSPORTATION. Slaughterhouses rarely take small numbers of chickens, but some do. Chickens are easily transported in cages. Several states allow sale of a certain number of home-processed birds.

BREED ASSOCIATIONS AND OTHER RESOURCES. The American Poultry Association and the American Bantam Association are two main resources for information. The Society for the Preservation of Poultry Antiquities is geared for heritage breeds, as well.

OTHER. Chickens can be useful allies in insect control. They scratch as they forage, which helps to break up manure pats of other species. This behavior helps to control parasites and flies.

CHICKENS EXPAND INTO NEW AREAS

Chickens are flourishing! Backyard flocks of hens are increasingly accepted in suburban areas and are experiencing a real boom in popularity. Nonetheless, check with your neighbors and local regulations before acquiring a flock, because chickens can be the target of some zoning restrictions and are not welcome everywhere. Given their noise level, this ban is especially the case for roosters.

Chicken breeder and author Pat Foreman promotes backyard chicken production through her books and public speaking engagements across the country. She has developed several strategies for incorporating chickens into gardening, as well as into suburban life. Her efforts have helped expand the numbers of people now keeping heritage poultry flocks and support new-owner education through her Gossamer Foundation.

Buckeye Chicken

Valued both as a broiler and for its medium-sized brown eggs, this dual-purpose heritage breed is on the increase and considered a conservation success story.

Homeland: Ohio

Traditional uses: Broiler and egg production

Size: Hens 6.5 pounds; roosters 9 pounds

Special adaptations: Medium-sized brown eggs. Cold-tolerant with pea comb and tight feathering

Colors: Red

Other traits: The most active of the American class chicken breeds

Conservation status: Threatened

Aylesbury Duck

Very rare now, the Aylesbury was esteemed and selected for its rapid growth rate. The skin is white, unlike that of most ducks.

Homeland: England

Colors: White with pale beige bill

Traditional uses: Meat

Conservation status: Critical

Size: Drake, 10 pounds; duck, 9 pounds

Ducks

DUCKS HAVE MANY OF THE ADVANTAGES of chickens: small size, docile character, prolificacy, and a short generation interval. These combine to make them a good choice for beginners learning the intricacies of animal management and selection. They are a useful, and often overlooked, species.

Size: Ranges from 1-pound bantams to 10-pound Rouens and Pekins

Average life span: 5 to 7 years

Handling ease: Easy and docile

Noise and odor level: The ducks (females) quack; the drakes (males) are nearly silent. Ducks are wet in their habits and can make a lot of mud and odor if not kept clean.

Shelter and space requirements: Hardy and resistant to wet and cold; they need shade in the hot summertime. Water is essential for health, at least enough to wash their heads and eyes in. A cage about 3 × 5 × 2 feet (1 × 1.5 × 0.6 m) is sufficient for 5 or so ducks of most breeds.

Zoning restrictions: Variable depending on location but usually few restrictions

DAILY FOOD AND WATER REQUIREMENTS. Duck feeds are widely available at feed stores, although somewhat less so than chicken feeds. Water is essential, and water for swimming and bathing is appreciated.

SOCIAL STRUCTURE. Ducks form groups of multiple ducks and a few drakes. Drakes can serve up to 10 ducks. When too many drakes are in a group they tend to worry the ducks incessantly.

REPRODUCTION. Specialized egg-producing breeds can lay an egg a day, which is more than most chickens! They mature at about 5 to 6 months old. Incubation is about 28 days. Larger breeds, such as the Rouen and Aylesbury, require a swimming area at least 6 inches deep so they can mate properly.

PREDATOR CONTROL. Essential. Tight housing can eliminate most predator threats, especially necessary at night.

PRODUCTS. Eggs, feathers, meat.

PROCESSING AND TRANSPORTATION. Small-scale slaughterhouses may be willing to process small numbers, but check your local facility for availability and pricing. Some states allow the sale of a set number of home-processed birds.

BREED ASSOCIATIONS AND OTHER RESOURCES. The American Poultry Association and the American Bantam Association are two main resources for information. The Society for the Preservation of Poultry Antiquities is geared for heritage breeds, as well.

OTHER. Ducks can be useful allies in slug, snail, and insect control. Some ducks are voracious consumers of flies.

DUCK FARMS

Ducks lay more eggs and are more vigorous than chickens, with fewer disease issues. Largely for these reasons, in the late 19th century and early 20th century, duck farms producing both meat and eggs were far more numerous than chicken farms in America, and massive flocks could be found throughout the country. But as American tastes shifted toward more refined and lighter fare, farmers shifted to more profitable chicken production, and the duck flocks declined.

Today, only a few massive flocks remain. Many duck breeds are now in desperate need of new stewards willing to rediscover the art and pleasure of keeping these remarkable birds.

Cayuga ducks

Geese

GEESE ARE LONG-LIVED AND LARGER than other poultry, with a unique character and products. They require a different sort of commitment than other poultry species and can be tricky for beginners. In the right place, though, they are a good choice for a conservation project.

Size: Shetlands, as small as 7 pounds; Toulouse, up to 26 pounds

Average life span: Up to 20 years or more

Handling ease: Moderate to difficult. Even with friendly individuals, geese can become highly stressed and will struggle when handled, so take great care when hands-on care is necessary.

Noise and odor level: Very high. Geese have been used as alarm animals because they make so much noise when startled or confronted with new situations. Minimal odor if kept on grass.

Shelter and space requirements: Geese are hardy in a wide range of weather conditions, so need only minimal shelter. They do not tolerate close confinement particularly well and prefer to range freely in large paddocks. They appreciate shade in hot weather.

Zoning restrictions: Common

DAILY FOOD AND WATER REQUIREMENTS. Mature geese are grazers, and grass can meet most of their needs. Young growing geese need a protein boost from prepared feeds, as do geese during egg production. Most geese do best with access to at least enough water to swim in occasionally.

SOCIAL STRUCTURE. Geese prefer to form pairs, although some breeds will form trios of two geese and a gander.

REPRODUCTION. Egg production is modest (35 to 60 per year for most breeds) compared with smaller poultry. Birds mature at about a year old. Incubation is 30 to 34 days. Larger breeds, such as the Dewlap Toulouse, require a swimming area at least 12 inches deep so they can mate properly.

PREDATOR CONTROL. The size of adult geese protects them from smaller predators, but larger ones (foxes and on up) are a threat. Goslings are susceptible just like ducks and chickens.

PRODUCTS. Meat, eggs (few), feathers, and down.

PROCESSING AND TRANSPORTATION. Transport geese in cages. Processing may be limited, as not all slaughterhouses are equipped to process geese.

BREED ASSOCIATIONS AND OTHER RESOURCES. The American Poultry Association is the main resource for information. The Society for the Preservation of Poultry Antiquities is geared for heritage breeds, as well.

OTHER. Geese are grazers and very selectively go after grasses. They are sometimes used to weed other crops.

Among the largest of geese, Toulouse geese tend to be slower and more quiet than other breeds.

Cotton Patch Goose

As its name suggests, the Cotton Patch goose has a long history of use in the southern United States as a valued ally in keeping grass weeds out of cotton crops.

Homeland: The Cotton Belt

Traditional uses: Weeding cotton fields; Christmas and Passover goose roasts

Size: Ganders, 14 pounds; geese, 12 pounds

Color: Ganders, white; geese, gray or gray saddleback

Conservation status: Critical

Bourbon Red Turkey

Its flavor, history, and foraging ability have made the Bourbon Red a breed of choice for farmers producing birds for Thanksgiving tables.

Homeland: Kentucky

Color: Red with white tail and wings

Traditional use: Thanksgiving roast turkey

Conservation status: Watch

Size: Toms, up to 33 pounds

Turkeys

TURKEYS ARE LARGE AT MATURITY but can forage well on their own in many situations. Annual production of turkeys for local Thanksgiving demand can be a great project to start learning about birds and management, easing into the longer commitment of managing and selecting breeding stock.

Size: Midget White hens, as little as 10 pounds; older Broad Breasted Bronze toms can reach 36 pounds

Average life span: Up to 10 years

Handling ease: Size on the larger ones can make them difficult to handle, but most turkeys become tame and tractable to handling.

Noise and odor level: Moderate. During the breeding season the toms gobble a great deal, but their call is usually not as piercing as the noise from roosters or ganders.

Shelter and space requirements: Adult turkeys are fairly resistant birds, so they do not need much shelter. Some option for shade in summer and a roof that sheds rain and snow in the winter is welcomed. They are large birds, so close confinement is usually not practiced.

Zoning restrictions: Variable

DAILY FOOD AND WATER REQUIREMENTS. Turkeys can forage for a great deal of their food. Poults need the extra protein that turkey and game bird feeds offer. Fresh clean water is essential for health.

SOCIAL STRUCTURE. A single tom can mate up to 10 hens. Multiple toms in a small group of hens can worry the hens and also fight among themselves.

REPRODUCTION. Sexually mature at a year. Turkeys lay eggs seasonally in spring and summer, and fewer eggs (up to 100 per year) than chickens. Incubation is 28 days. Very young poults need careful attention during brooding. Egg production in turkeys is related to size of the hen. The lightest hens lay 85 to 100 eggs per year, medium-weight hens produce 50 to 75, and heavy hens about 50.

PREDATOR CONTROL. Mature birds are resistant to all but large predators. Growing poults are very susceptible to a wide variety of mammalian and avian foes.

PRODUCTS. Meat, feathers, rarely eggs.

PROCESSING AND TRANSPORTATION. Transportation is by cages. Processing is variable at local slaughterhouses, so be sure to check in advance.

BREED ASSOCIATIONS AND OTHER RESOURCES. The American Poultry Association is the main resource for information. The Society for the Preservation of Poultry Antiquities is geared for heritage breeds, as well.

OTHER. Turkeys have long been valued for their ability to forage for most of their food, helping to control insect pests in the process.

SMALL BIRDS A BIG HOLIDAY HIT

Victoria and David Miller of Canyon Creek Farms of Sequim, Washington, have been raising and producing Midget White turkeys on their homestead for the local holiday market. A processing plant is not nearby, so they came to depend on another strategy to get the birds ready for customers. When the birds are ready for slaughter the Millers reach out to the community for help on processing day. They were surprised to find that all of their customers were interested in learning how to dress out their own holiday bird.

For their customers who want to try their hand at raising birds themselves, the Millers are now raising poults until they are fully feathered, 8 to 10 weeks of age, and ready to be put outside on pasture. This has created an excellent opportunity for inexperienced owners and a new market for the Millers.

Sheep

AMONG THE SMALLEST of the mammalian livestock, sheep are successfully raised by all ages and thus can be a good project for beginners learning livestock management and selection. Their multiple products appeal to a wide range of consumers and allow broad variation in conservation projects.

Size: Smallest Shetlands or Southdowns are 50 pounds or so. The largest Suffolk rams are up to 400 pounds. Ewes of most breeds are 90 to 150 pounds, rams about 200 pounds.

Average life span: Up to 14 years old is fairly routine, but production declines in most individuals after about 8 to 10 years old.

Handling ease: Larger sheep can be somewhat difficult to manage, depending on temperament. Most are easily managed. Rams are potentially dangerous, and caution should always be used around them.

Noise level: Moderate. They can be noisy if separated from the rest of the flock.

Shelter and space requirements: Shelter from cold and wet is minimal, but shade in hot sun is essential. For improved pasture the general rule is 5 sheep per acre.

Zoning restrictions: Variable, but generally widely acceptable in rural areas.

DAILY FOOD AND WATER REQUIREMENTS. Most breeds obtain all their requirements from grazing. During pregnancy and lactation, some concentrated feeds may be needed. Water needs vary with breed. In some seasons sheep may not drink at all, but clean water should always be available.

St. Croix sheep are raised for meat. See page 93 for a Breed Snapshot.

SOCIAL STRUCTURE. Sheep are social, and the general recommendation is that they are unhappy if fewer than three are in the flock. One ram can serve about 35 ewes reasonably well.

REPRODUCTION. Many ewes will mature at 6 months, ready to mate their first fall. Some breeders will wait to mate them at 18 months, for optimal growth and size. Many breeds will mate only in the fall, ensuring spring lambs. Others will mate year-round. Ewes typically produce one or two lambs, but triplets are routine in some breeds. Gestation is about 5 months.

PREDATOR CONTROL. Sheep are very vulnerable to predators, so some strategy, such as use of a livestock guardian, will be needed to avoid losses.

PRODUCTS. Meat, wool, hides, milk. Specialty markets for hand-spinning wool can be quite lucrative, but require extra care and labor in matching the fleeces to the customer.

PROCESSING AND TRANSPORTATION. Most slaughterhouses will process lambs. Transportation is usually on the back of a pickup truck in a cage, or on a trailer.

BREED ASSOCIATIONS AND OTHER RESOURCES. Each breed has its own breed association, but a more general starting point is the American Sheep Industry Association. Information on sheep management is widely available from most county extension offices within each given state.

OTHER. Foot-trimming is a specialized skill that can be readily learned and is essential for overall health and well-being of the sheep. Shearing is another specialty skill, and good shearers can be difficult to find in some areas. Poor shearing can ruin a year's growth of wool in seconds.

NEW MARKETS FOR FLEECE AND FIBER

The electronic web-based information age has been a great boon to fiber producers as well as their customers. Many producers of wool from heritage sheep breeds have created a strong and lasting demand for fleeces and further processed fiber by creating private websites to promote and sell their products. Many are also taking advantage of more central sites such as Ravelry, craigslist, eBay, and Etsy. Being able to make direct sales to a national market through the Internet tends to bring a higher price to the producer, and be a bargain for the consumer. Everybody wins!

Romeldale/CVM Sheep

The wool of this breed is very soft and fine and comes in several colors. For years it was all sold to the Pendleton Woolen Mill. The Romeldale is the white version of this breed, while CVM stands for California Variegated Mutant, one of several color patterns that also occur in the breed and lead to a wide array of fleece colors.

Homeland: USA

Traditional uses: Fine white and colored wool; meat

Size: Rams, 275 pounds; ewes, 150 pounds

Colors: White, grey badgerface, black, and moorit brown

Conservation status: Critical

St. Croix Sheep

St. Croix are hair sheep (producing no wool) and well adapted to hot, humid climates. They are prolific, nonseasonal breeders. The sheep are polled (naturally hornless).

Homeland: St. Croix, U.S. Virgin Islands

Traditional uses: Meat

Size: Rams, 165 pounds; ewes, 120 pounds

Colors: White; other colors, rarely

Conservation status: Threatened

Goats

MOST GOATS ARE SMALL and easily handled, but larger, horned animals can be tricky. Their curious and adventurous ways can make management a challenge, so beginners need to anticipate potential weak spots in facilities or abilities. They are charming, but impish!

Size: The smallest does, about 50 pounds; the largest bucks, about 300 pounds, sometimes even larger

Average lifespan: 10 years is routine; a few reach 15 years

Handling ease: Moderate to easy. Wilder and larger goats present a challenge. Bucks can be a problem, especially if completely imprinted on people, but are rarely the threat that rams can be.

Noise and odor level: Moderate for noise. Odor is variable, with bucks in the fall breeding season being especially aromatic. This pungent smell is one reason for widespread zoning restrictions against goats.

Shelter and space requirements: Goats do not like to be wet, and so some sort of roof to allow them to get out of rain is appreciated. They are otherwise fairly resistant to weather. The usual recommendation for improved pasture is 5 goats per acre. Most goats climb incredibly well, and this must be anticipated in housing and fencing them.

Zoning restrictions: Common

DAILY FOOD AND WATER REQUIREMENTS. Grazing and browsing (reaching up for leaves) should provide most needs for goats. They need concentrates if used for milk production and also during late pregnancy. Feeds are readily available at feed stores.

SOCIAL STRUCTURE. Goats are social and prefer to be in groups of at least three animals. Males will fight during the fall breeding season. Does jostle for position routinely, so some level of discord is common in goat herds.

REPRODUCTION. Does of most breeds will mate only in the fall, for the arrival of spring kids, but this cycle can be variable and many heritage breed does can mate year-round. Some breeds mature in their first fall season (6 months old), others in the next fall (18 months old). Gestation is about 5 months. Most does produce one or two kids, but litters of three, four, or even five do occur.

PREDATOR CONTROL. Essential. Goats need some sort of protection from predators or there will be losses.

PRODUCTS. Meat, milk, fiber (cashmere, mohair).

PROCESSING AND TRANSPORTATION. Most slaughterhouses will process goats. Goats are usually hauled in cages on the back of pickup trucks or on trailers.

BREED ASSOCIATIONS AND OTHER RESOURCES. Each breed has its own registry association, but broader information can be found at sites such as the American Dairy Goat Association.

OTHER. Specialty goat cheeses are in high demand in some areas. Goat manure is highly sought by gardeners.

EACH BREED HAS ITS ADVANTAGES

One of the unique benefits of the Myotonic (Tennessee Fainting) goats is their inability to climb or jump. This, in addition to their quiet temperaments, can make them acceptable where other more adventurous or vocal goats might be unwelcome. You might ask yourself "Why would I want a goat that faints?" Four very good answers have kept this breed popular with producers.

First, they do not actually faint. These goats have a genetic condition called *Myotonia congenita* that causes a stiffening of muscles whenever the animals move quickly, especially when startled or excited. When severely startled, the legs will stiffen and cause the animals to, on occasion, fall over. Mind you, this in no way hurts or disturbs the animals and they often continue chewing their cud as their legs un-stiffen! This condition acts on their bodies like isometric exercise and causes good muscle development and excellent meat-to-bone ratio.

Second, the stiffening of the legs makes the animals less agile than goats of other breeds.

They are not able to jump fences as easily as other goats. (If you have ever encountered the challenge of trying to contain a goat, you will truly appreciate this trait!)

Third, they are quiet. Their quiet habits endear them to many breeders, and no doubt to their neighbors as well!

Fourth, their efficiency makes them easy to maintain and enjoy. Their efficiency includes good reproductive rates, good use of feeds and forages, and also good mothering ability.

Myotonic (Tennessee Fainting) Goat

This breed's parasite resistance, good mothering ability, and other traits have made it a conservation success story.

Homeland: Tennessee and Texas

Traditional uses: Meat production, brush control

Size: Does, 60–125 pounds; bucks, 80–200 pounds

Colors: Usually black and white, but colors vary widely

Other traits: Some are shaggy, some are smooth, some are horned, some are polled

Conservation status: Recovering

Gloucestershire Old Spots Hog

The mothering ability of sows of this breed makes them enormously valuable to producers, since all the offspring are likely to be healthy and survive to maturity.

Homeland: England

Traditional use: Pork production on grass and in orchards

Size: 500–600 pounds

Color and other traits: White with black spots

Conservation status: Critical

Pigs

THE CHALLENGES OF PIG MANAGEMENT and care require at least some caution in choosing swine as a first project. Nevertheless, their litters and generation interval provide good scope for breeders to practice selection and watch as it affects the herd.

Size: The smallest Guinea Hog sows are about 100 pounds. The largest Red Wattle boars are nearly 1,200 pounds!

Average life span: 6 to 8 years, but usually much less than this in commercial settings. Jeannette Beranger worked with a Guinea Hog sow that reached the ripe old age of 15 years.

Handling ease: Most hogs are relatively easy to handle, but variation occurs and some hogs are a real handful. Some boars, of some breeds, can be aggressive and dangerous. Sows with litters can also be fiercely protective.

Noise and odor level: Moderate in day-to-day activities. Piercingly loud squealing accompanies restraint or uncertainty, and if it often occurs, neighbors may object. Moderate odor if kept clean, overbearing if not.

Shelter and space requirements: Hogs require at least simple housing, such as an A-frame structure, to get out of cold, wet weather and have a dry place to sleep. They also need shade and a wallow to help them keep cool in the summer. Wallowing and rooting can cause quite a bit of damage if done in the wrong place.

Zoning restrictions: Common

DAILY FOOD AND WATER REQUIREMENTS. Hogs can be raised on pasture, but they usually also require at least some sort of provided concentrated feed or kitchen scraps. Clean water for drinking is essential, and water for wallowing is essential in hot weather.

SOCIAL STRUCTURE. Sows develop a strong social order. Boars are usually housed separately and brought out only for mating with the sows, or for use in a small group of sows as a breeding group. Boars will fight among themselves.

REPRODUCTION. Mating occurs year-round, and gestation is just short of 4 months long. Most animals mature at about 6 to 8 months, but most breeders wait until gilts are at least 8 months old to ensure good litter sizes. Heat cycles occur every 21 days. Litter sizes vary widely both by breed and by individual sow.

PREDATOR CONTROL. Hogs are more capable of self-defense than are sheep or goats, but in some regions predators must be considered.

PRODUCT. Meat.

PROCESSING AND TRANSPORTATION. Be sure to check that your local slaughterhouse accepts swine, because not all are equipped (or willing) to process hogs. Most will not process intact boars. Transportation is usually in the back of a pickup truck or on a trailer.

BREED ASSOCIATIONS AND OTHER RESOURCES. Breed associations are available for each breed.

OTHER. Some farmers have success in using hogs for working the soil for later planting of crops or trees. The rooting and scavenging can rework the landscape dramatically. In the deep South they have been used for snake control.

MATCH THE HOG TO THE SITUATION

Hog breeds vary widely, and a hog can be found that fits nearly every situation. The large, lop-eared breeds (such as the Gloucestershire Old Spots) tend to be the most docile and easy to manage but they can become very large. The breeds with more erect ears, or the hogs with a recent feral background (such as the Ossabaw Island hog), can be more aggressive but are better choices if extensive ranging in woods is the desired management system. Hogs also vary widely in temperament depending on the breeders' preferences. In the case of the Ossabaw, you may find some that are very docile and friendly while other bloodlines are very self-sufficient and aggressive. Each has advantages depending on your needs and farming strategies.

Keep in mind, however, that in all cases, no matter what the reputation of the breed or individual animal, sows can be defensive of young, so plan facilities and management accordingly. The goal is safe animals and safe people.

Red Wattle pigs

Cattle

CATTLE ARE LARGE, STRONG, and long-lived. They need space and a long-term commitment to their breeding and care. While breeders with a wide range of experience have successfully raised cattle, beginners must start by carefully researching their care and management.

Size: The smallest cows are 500 pounds or so, and large bulls are 3,000 pounds or larger. Most heritage cows are from 800 to 1,200 pounds, depending on breed.

Average life span: Most cows live about 10 years, with some breeds exceeding this by quite a bit and living to 20 years or beyond.

Handling ease: Variable. Range cattle can be difficult or dangerous due to their unfamiliarity with people. They need to be worked through strong chutes to keep them and their handlers safe. Dairy, and other cattle that are handled daily, may well be halter-broken and very easy to manage.

Noise and odor level: Moderate. The lowing of cattle is rarely problematic except at weaning when cows call loudly to their calves.

Shelter and space requirements: Cattle tend to need minimal shelter, and many cattle never see the inside of a building their entire lives. Cattle are large and need space. The usual recommendation is a cow per 2 acres of improved pasture.

Zoning restrictions: Zoning in many residential areas prohibits cattle.

DAILY FOOD AND WATER REQUIREMENTS. Beef cattle are usually left to forage and graze on their own with little supplementation. Dairy cattle typically need a boost from some concentrated feeds. Water is essential for cattle, and dairy cattle need to consume up to 40 gallons a day.

SOCIAL STRUCTURE. Cows are herd animals and prefer the company of herd-mates but are usually less adamant in this than are sheep and goats.

REPRODUCTION. Cows cycle year-round and have a 21-day estrous cycle. A single calf is produced after a gestation of a little more than 9 months. Many breeds mature at about a year old, allowing them to conceive and then calve when they are reaching their second birthday.

PREDATOR CONTROL. Small predators are not a problem, but situations vary and larger predators can cause problems.

PRODUCTS. Beef, milk, hides, labor from oxen.

PROCESSING AND TRANSPORTATION. Slaughter facilities for cattle are routine throughout much of the country. Transportation of such large animals can be tricky, and most producers rely on trailers.

BREED ASSOCIATIONS AND OTHER RESOURCES. Each breed has its own association.

OTHER. Several heritage breeds excel at ox production. In some regions of the country the demand can be quite high for steer calves that can be trained to work as oxen.

LOOKS CAN DECEIVE

Breeds with long impressive horns all look aggressive and formidable. Some, like the Texas Longhorn, are actually docile and generally easily managed. Others, like the Ancient White Park, retain some of their feral heritage in a wild streak that requires good facilities and knowledgeable managers in order to keep both cattle and handlers safe at all times.

Ancient White Park

Horses

HORSES DO NOT PRODUCE EDIBLE or wearable products — they are valued for their performance, and this usually demands training. Trained horses are valuable partners; untrained horses can be a hazard. Beginners must take steps to ensure the training and safety of both animals and people.

Size: The smallest are about 10 hands (40 inches) in height and 400 pounds; the largest are 19 hands (190 inches) and close to 3,000 pounds.

Average life span: 20 years is routine, and up to 30 is relatively common

Handling ease: If trained, easy. If untrained, challenging and dangerous

Noise and odor level: Low

Shelter and space requirements: While horses can stay out in just about all kinds of weather, many people provide them at least a three-sided run-in shed.

Zoning restrictions: Few

DAILY FOOD AND WATER REQUIREMENTS. Grass or good hay is adequate for most horses, but those in active work need concentrates or grain.

SOCIAL STRUCTURE. Horses are herd animals and prefer the company of a few well-known companions. The happiest herds result from mares mixed with mares or geldings with geldings, as geldings can be rough on mares.

REPRODUCTION. Mating is usually during the spring or summer, but some can breed year-round. Gestation is about 11 months, and foals (one per mare per year) are usually born in the spring. Mares' estrous cycle is 21 days at the peak of the season. Fillies are usually first mated at about 3 years old.

PREDATOR CONTROL. Only the largest predators succeed against horses, mountain lions being the most successful. This is not a concern for most owners.

PRODUCTS. Riding, draft

PROCESSING AND TRANSPORTATION. Processing is a non-issue for this species. Transportation is usually by trailer.

BREED ASSOCIATIONS AND OTHER RESOURCES. Each horse breed has its own association.

OTHER. A great deal of the value of horses is in what they can do. This comes from training, and good training can be the difference between a very valuable animal and one with marginal to no value.

PERFORMANCE HORSES BUILD DEMAND

Active campaigning of Dales Ponies in competitive driving by Pat Hastings of the Hamilton Rare Breeds Foundation (Hartland, Vermont) has enhanced demand for a breed that needs all of the promotional help it can get to keep numbers viable. Hamilton Rare Breeds Foundation has had great success in the show ring with this breed.

Donkeys (Asses)

DONKEYS HAVE THE SAME ISSUES as horses: their products are not edible, and training is essential to keep animals and people safe. Donkeys have the added asset (or liability) of that great voice! Beginners must carefully assess their ability to manage asses successfully.

Size: Miniatures, as short as 28 inches (71 cm); Mammoths, up to 64 inches (163 cm) and even taller

Average life span: Tend to live into their 30s

Handling ease: Easy if trained, hazardous if untrained. Jacks are potentially dangerous.

Noise and odor level: High, especially for jacks braying in the mating season

Shelter and space requirements: Donkeys are more cold-sensitive than horses are and need to be able to get out of cold or wet weather.

Zoning restrictions: Usually similar to horses

DAILY FOOD AND WATER REQUIREMENTS. Donkeys are generally very easy keepers, and grass and hay are usually sufficient; in fact, they are prone to obesity if overfed. They are efficient with water, but do need a constant source of clean drinking water.

American Mammoth Jackstock

SOCIAL STRUCTURE. Donkeys prefer the company of other donkeys and do best when at least two are together. Jacks are more aggressive and do well when alone.

REPRODUCTION. Donkey gestation is longer than that of horses and usually lasts for a full year, resulting in a single foal. The estrous cycle is 21 days, and jennies are usually first mated at 3 years old.

PREDATOR CONTROL. Donkeys are used for predator control by many producers of small stock, and are the targets for only the largest and most capable predators such as mountain lions. They have an innate distrust for dogs, coyotes, and wolves, and usually go after them.

PRODUCTS. Riding, driving, predator protection.

PROCESSING AND TRANSPORTATION. Processing is not an issue, and transportation is usually by trailer.

BREED ASSOCIATIONS AND OTHER RESOURCES. The American Donkey and Mule Association is the main source of information on donkeys. Mammoth Jackstock owners are served by their own independent association.

OTHER. Donkeys' dislike for predators is used to good advantage by many owners of small livestock. Not all donkeys will do the job of guarding livestock, but the good ones are great at it.

REVERSING THE SLIDE

At the Hamilton Rare Breeds Foundation in Hartland, Vermont, Debbie Hamilton has been instrumental in reversing the slide to extinction of the Poitou donkey. This large mule-breeding

donkey hails from France, where numbers had become disastrously low. The big, attractive animals need special care in their first few days, and the skills to provide this are no longer common. Debbie has not only been able to ensure that the foals get a good start, but her excellent eye for stock and for breeding has achieved high-quality animals. As a result she has been able to coordinate closely with the French in saving this unique heritage breed.

Buyer beware: Not all shaggy donkeys are Poitous! Only individuals with French registration papers can be considered purebred Poitou.

CHAPTER FOUR

Getting Started with Heritage Breeds

So, YOU HAVE DECIDED WHICH SPECIES will thrive on the land and climate of your farm, under your preferred type of management. You have considered what you are going to produce (meat, milk, eggs, riding or draft animals, or pets), and how you will process and sell those products. Now it's time to choose, from the wonderful array of breeds, the one that is right for you and to get started.

Choosing a Breed

CHOOSING A BREED TO WORK WITH is often the most rewarding and fun part of getting involved with heritage breeds. Linking your own interests, abilities, and facilities with the needs and status of the breed is exciting. Making sure the fit is good all the way around benefits the farmer, the breed, and the future of agriculture.

The most successful conservation projects come from working with a specific breed that holds great interest for you. When well matched, the joys of managing the animals far outweigh any of the occasional inevitable frustrations. Take time to investigate breed choices for your goals and interests, and choose the one that has special appeal for you.

If you are still undecided about which project to undertake, another approach is to choose a breed that will fit your farm, and then investigate which projects will help that breed the most. Some breeds have endangered bloodlines, and a few dedicated producers working with these would do wonders to make certain the bloodlines do not disappear.

DECISIONS FOR A RARE BREED PROJECT

Here are some things to think about as you embark on a project raising heritage breed animals.

1. What do you want to do?
Meat? Eggs? Milk or cheese? Draft animals? Fiber? Riding animals?

2. What species will meet your needs?
Sometimes species other than the obvious can do the same job.

3. Do you have the proper infrastructure to manage the species?
This includes housing, fencing, water, feed or pasture, handling equipment, veterinary support, and ability to move animals off the property in case of emergency.

4. Do you have or can you acquire the skills to manage the species?
Some species and breeds are more challenging than others.

5. Which breeds will do the job?
Multiple breeds might meet your goals.

6. Which of those breeds thrives in your climate and landscape?
Are they cold-adapted? Heat-tolerant? Best kept in wet or dry climates? Resistant to local parasites?

7. Do you like the breed?
Once the choices are narrowed down, the last question you ask yourself is "Which one breed do I really like best?"

Take some time to pore over the breed-by-breed information in the Appendix at the back of this book, beginning on page 188. Next, check out the comprehensive breed listings on the The Livestock Conservancy website (see Resources). Once you have spotted some likely candidate breeds, your next task is to find out if animals and resources are available locally, or at least regionally. Remember, these are rare breeds. Having a source for your initial stock close at hand keeps shipping or trucking costs at a more reasonable level, not to mention that having other breeders close by can serve as part of your support system. If you are firmly committed to a particular breed, no matter the obstacles, you may have to travel great distances to obtain quality stock.

To find breeders and other resources, contact the breed association. If possible, follow up by visiting with several breeders. This will help you become familiar with the animals, the people who work with them, and verify whether they are truly a good fit for you and your farm. Ask lots of questions. Why this breed? What do you like best about it? What do you like least about it? Especially observe and ask about temperament, because in your day-to-day management it will be much more important to be happy with your animals' dispositions and attitudes than with their looks. Meeting breeders also ensures that a prospective buyer will be able to develop an appreciation for their varied approaches to animal management, production, and marketing. These visits will help you to refine the goals for your project.

Finding heritage breeding stock, whether to start or expand a herd or flock, is usually an ongoing challenge for heritage breeders. Consequently, finding the desired combination of bloodline, quality, temperament, and adaptations takes time, especially as you become more expert with the breed and your desires become more specific. Patient breeders will recognize that what they want may not be located nearby, and that breeders of the highest-quality stock often have waiting lists. Good relationships with the breed association and active breeders who are themselves well connected can ease the process of finding breeding stock.

Providing the Basics for Your Animals

ALL ANIMALS NEED SHELTER, SPACE, FOOD, water, and security. They also need good veterinary care when they are ill or injured. Your ability to provide these, along with the cost of building and maintaining facilities and providing feed and care, should be an important part of your decision-making process when choosing a species. To help you in this process, this section covers the basic needs of all livestock and poultry, as well as some of the specific needs of the individual species.

Getting Ready for Your New Arrivals

A barn or barns, well-planned pens, feeders, and waterers are essential for both animal and human safety and will greatly ease your daily chores. Avoid the mistake of investing in top-quality animals and then housing them in poor facilities that can result in accidents or ill health. Making routine care as easy as possible will also leave you with extra time and energy to enjoy your animals and to observe and reflect, which are so important for making wise breeding decisions. Above all, plan to have your infrastructure in place before the animals arrive!

That said, facilities do not need to be state-of-the-art (and in many cases should not be, at least for heritage livestock). Advanced materials and construction that characterize state-of-the-art modern facilities usually have high maintenance costs that go along with them. More traditional facilities are usually cheaper to build and maintain, and in most cases adequately support the shelter needs of heritage breeds. Cleanliness and safety should always be primary considerations. Clean and safe usually equals fun, and when projects are fun they are likely to continue for a long time.

Each species and each region has different facility needs, so the general recommendations in this chapter should always be supplemented by local information. Check with local extension agents, or other breeders, to find out what works in your area, and follow their lead while always thinking of innovations you could add to make routine care as easy and safe as possible.

What Animals Need in Order to Thrive

Each species has specific needs for space and environment. Ideally these should match up with what is available and affordable for you as an owner. The bottom line is that it is not hard to take *really* good care of your animals if you keep in mind the following basics when planning a rare breed project.

AS THE SPECIES GET LARGER AND STRONGER, SO MUST THE INFRASTRUCTURE NEEDED TO MAINTAIN THEM. Bigger animals require larger structures, more space, and generally a bigger budget.

EVERY ANIMAL SHOULD HAVE THE OPTION OF GETTING *IN*, OUT OF THE WEATHER. The type and quality of housing needed depends on the species, natural environment, and climate. For warmer climates, closed barns may not be necessary and perhaps simple run-in sheds are all that is needed for most species. But every animal should have the option to get out of weather extremes (even if the more hardy ones choose not to). This includes extreme heat, bitter cold, storms, strong wind, or other adverse weather. Sometimes a simple windbreak such as a hill or tree line will do for hardier animals. More sensitive species may need a shelter or barn.

EVERY ANIMAL SHOULD HAVE THE OPTION OF GETTING *OUT*, INTO THE WEATHER. Never underestimate the power of fresh air and sunshine on a daily basis. Even if exposure is for only part of a day, being outside is stimulating and will contribute greatly to the well-being of your animals. Animals are generally happiest and healthiest with full access to the outdoors—whether they use it or not.

In addition to these general rules, here are a few more basics.

Environment

Animals need an environment in which they are comfortable, clean, secure, and have room to roam for exercise and stimulation. The area where they are expected to spend the majority of their time should make them feel at ease, keep stress to a minimum, and provide stimulation that combats boredom. As a minimum they should have companions and be able to see beyond the enclosure. They should also have access to dirt or forage. Specific recommendations are highly variable depending on species, breed, and location, so check locally to see what works in your area. Animals that are left by their owners in a cage or stall all day, day in and day out, can be frustrated, and that can lead to aggression or self-destructive behaviors that can be difficult to correct.

Security

We all want to keep our animals protected and secure, which means keeping them in where they are supposed to be and keeping predators out. Depending on the species, this task is accomplished with a range of strategies from small cages to miles of fence line. Understanding the strength and capabilities of your animals to climb, dig, push over, or wiggle through pen walls and fences, and knowing what types of fences, barriers, or enclosures are successfully used by other owners, is vital. Fencing choice is particularly important with potentially destructive or aggressive animals, such as breeding bulls or stallions. Building strong, safe facilities can be a big expense, but cutting corners in this area is a recipe for disaster.

For more vulnerable animals, such as sheep, goats, and poultry, security may include the addition of guardian animals, such as dogs or donkeys to protect against large predators. For even smaller species and younger animals, such as chicks, poults, or young rabbits, the caging needs to effectively keep out smaller predators, such as snakes or rats.

Being able to safely move animals on and off the property is also important, not only for routine buying, selling, and breeding but also in case of an emergency such as a fire. With smaller species, a simple pet carrier or cage will do the trick; but with larger animals, how do you move animals on and off the property safely? Some breeders of large livestock resort to hiring livestock haulers on those rare occasions when livestock must be moved on or off the farm. Others prefer to have their own trailers, and this certainly has advantages if emergency evacuation becomes necessary.

And last but not least, if your animals get out of their enclosure, how will you get them back? Some can be trained to come when called, but that

SMALL STOCK MUST BE SAFE FROM PREDATORS

Predator control is important for all poultry species and for smaller livestock such as sheep and goats. Predator threats range from birds (hawks and owls) to mammals (weasels, coyotes, and raccoons, up to mountain lions and bears). The type and number of predators vary from place to place, so predator control practices vary as well. No single strategy works for all situations, but effective strategies are indeed available in all areas. Ask other producers what types and combinations of facilities (fencing, housing) and guardian animals (donkeys, llamas, dogs) are effective for them. Janet Vorwald Dohner's book *Livestock Guardians* (Storey Publishing) is a very useful guide to many of the options available.

does not always work with a panicked animal in a strange environment. Some folks resort to herding dogs. Be sure to lay out your fields with regard to how animals think: for example, they are more likely to see an open gate in a corner (and go through it!) than they are a gate in the middle of a side of a field.

Food and Water

All animals need access to sufficient (and clean) food and water. In fact, nothing is more important than access to clean water, and plenty of it. Do you have enough water resources on your property to manage the animals you plan to purchase? Water needs for the different species are highly variable. Feeds are also variable, species to species, and both water and feed requirements can be found in the National Academy of Science guides for the species. If your system is pasture-based, do you have enough acreage and is your pasture quality up to the task of providing good diverse nutrition for your animals? If you will be purchasing some or all of the feed you will need, can you afford your future feed bill?

Social Structure

People often overlook the behavioral and other management problems that arise from improper social structure. Almost never would domestic livestock or poultry choose to lead completely solitary lives, and a correct social structure for the species can be vital for reducing stress or encouraging successful breeding. When you keep multiple animals in a herd or flock, take care that the sex ratio is appropriate. If not, females (especially poultry) can be mated by the males too often for their health and well-being. Are too many males for the group leading to aggression and fighting? Alternatively, if you cannot keep more than one animal on your property, then you as the owner have the responsibility of providing stimulating and positive company for your animal. Most introductory books on animal science and animal husbandry have useful chapters on the social structure of a given species. Keeping a solitary animal happy and content can be a challenge, and housing just one animal is generally not recommended.

Veterinary Care

A good veterinarian who can handle livestock and poultry is almost as rare in some areas as the endangered breeds you want to work with. If you plan to actively raise and breed animals, a highly competent vet will be vital to your program, especially for those species that required a considerable investment of capital to bring on to your property, such as pigs, horses, sheep, or cattle. Check with others in your area who keep the species you intend to work with and see if they can provide recommendations for effective and compassionate veterinary care.

You have to do your part to maintain your animals' health as well. Understanding the basic veterinary needs of your species, such as annual vaccinations and preventative care, is a must. Some states have requirements for testing or vaccination of your animals, so check with local regulators to see what is needed. Regulators are usually a county extension service, or most states have state veterinary officers who are available to guide you through the requirements. Good local veterinarians will also have this information. If you plan to move animals across state lines, you will need to know the animal import requirements of the destination state so that your animals can travel legally.

American Mammoth Jackstock donkey

EVERY ANIMAL DOES NOT FIT IN EQUALLY WELL EVERYWHERE!

Smaller livestock and poultry tend to require much less space and a simpler infrastructure than do the larger animals, so these species are often the easiest projects to start with, particularly when undertaking the project on a part-time basis rather than as full-time farming. Rabbits and poultry have some breed choices that are great for children and for adults with limited space and facilities.

Keep in mind also that sometimes breeds within species can vary in their requirements as much as those across different species. An enormous Mammoth Jackstock donkey housed in a small backyard would likely end up annoying neighbors with its braying and would pose a host of challenges that could be avoided by deciding to get a Miniature Donkey that requires much less space. But the braying may still annoy the neighbors!

Steps to Get Underway

WHAT DO YOU STILL NEED TO LEARN AND DO before starting your heritage breed project? A few specific steps remain, including:
- setting realistic goals and budgeting to meet them
- animal identification and record keeping
- planning for the marketing of production and excess breeding animals
- knowing your exit strategy (see page 185)

But first, and most important, you should look at what you expect from, and can realistically give to, a breed.

American Mammoth Jackstock Donkey

This breed, the largest among donkeys, was specially adapted for mule production — raising jacks to be mated to mare horses. Their large conformation was developed specifically for the production of large draft mules.

Homeland: Middle and Deep South

Traditional use: Mule production

Size: Jennies, over 14 hands (56 inches); jacks, over 14.2 hands (58 inches)

Colors: The historic color was black. Color is more variable now, with odd non-black colors bringing a premium.

Other traits: Conformation was for draft use originally. Modern-day uses include riding.

Other: These animals can be very noisy, especially jacks in the breeding season. Jacks can be aggressive to other donkeys, as well as to people, but their behavior is variable.

Conservation status: Threatened

Start with Realistic Goals

Successful heritage breed projects have realistic goals that focus and drive the project. Matching your interests to your available time, budget, experience, facilities, and infrastructure is an important part of deciding on your goals. Be careful not to bite off more than you can chew, because overwhelmed breeders rarely continue to be successful in their projects. A modest project that is well executed will benefit the breed (and the farmer) far more than an overly ambitious project that is abandoned partway through due to frustration.

Some small-scale projects with rare breeds have an impact that goes beyond their modest size. Elaine Shirley, of Colonial Williamsburg, has worked to multiply their flock of Nankin bantams and has successfully figured out how to ship hatching eggs across the country so that others can easily start with this small, appealing breed.

See page 116 for a Breed Snapshot of Nankin bantams.

Matching Goals, Abilities, Preferences, and Budget

EACH SPECIES OF LIVESTOCK AND POULTRY has its own unique combination of physical and behavioral characteristics. Previously we discussed the basics of what each species needs for food, water, shelter, and space. Using this information, take the time to figure out which type of livestock or poultry most fits your budget, space, facilities, abilities, available time, and interests. The wrong animals in the wrong place can end up being a management nightmare for both the animals and the caretaker. Carefully consider whether you can take care of their needs *before* embarking on a project, and you will make their lives happier — and yours as well.

A vivid example of "getting in over your head" is portrayed in the documentary movie *Buck* about the horse trainer Buck Brannaman. In that film one well-meaning woman has a rogue young stallion that was bottle-raised and as a result lost all fear of and respect for people. The horse had turned into a monster looking for people to dominate and hurt — and was one of several intact stallions the unfortunate woman now had. The situation had grown from a manageable one into too many animals of the wrong type, in the wrong place, with inadequate opportunity to manage any of them for a good final outcome. These are avoidable mistakes!

Next, before getting too far down the road of choosing a breed within your chosen species, reflect on what you want to produce or accomplish, and your preferred way of doing things when working with animals. Many breeders come to breed conservation projects with some goals already firmly

Nankin Bantam

Tiny size and a strong brooding and maternal instinct are distinctive traits of this unique breed. Present in England by the 1500s, the Nankin bantam may have originated in Southeast Asia.

Homeland: Exact origin unknown

Traditional uses: Eggs, exhibition, broody hens for incubating game birds and other eggs

Size: 22–24 ounces

Color and other traits: Red gold with a single or rose comb

Conservation status: Critical

single comb

rose comb

in place (for example, wool versus meat for sheep, eggs versus broilers for chickens, dairy versus beef for cattle), which helpfully whittles down the list of candidate breeds. Knowing what you want to do and how you want to do it makes it simpler to choose an appropriate breed.

For example, meat production will vary depending on the species and the farm system. For cattle, production choices include entirely grass-based, or corn finishing (both of which are traditional, heritage practices in different regions). The end product can range from young animals processed for veal (milk fed) or rose veal (milk- and pasture-fed older calves) on up to products from cattle of varying ages and sizes. If you plan to finish and sell the animals yourself, doing some market research ahead of time is also important to determine whether the product you want to produce can be readily sold in your area. Each combination of goals (grass-based rose veal, grass-finished beef, corn-finished mature beef) will likely entail a different breed choice.

Personal preferences for animal handling should also figure into your breed choice. Some breeders prefer a "hands on" approach with close management, careful feeding, and great attention to individual animals. Others are more interested in large populations or in animals that go out and fend for themselves for nearly everything they need. For nearly every strategy for animal production, a breed can be found that is a good fit.

Specialized products require capable husbandry as well as talented processing and promotion. Frank Reese has had successful chicken, turkey, goose, and duck production as a result of his talented selection and management. In addition, he has located a processor that helps him prepare the final product so that it meets the demands of discriminating consumers.

Budget

Your budget for starting is an important "detail" that can sink an otherwise well-intentioned plan. Effective projects with rabbits, chickens, ducks, turkeys, and geese can often be started with a few hundred dollars, and in some cases less than that. These species are also relatively inexpensive to maintain. At the other extreme are performance horses, which might have a starting price tag of several thousand dollars, to which must be added the costs of grain, hay, a trailer (and a truck to pull it), tack and harness, farrier (horse shoer/hoof trimmer) services, and routine health care. Soon the initial cost of a horse pales in comparison with adequate care and upkeep.

Finding a Balance

Plenty of worthwhile projects have budgets between the two extremes of performance horses and backyard poultry. Keep in mind that the initial cost of the animals is usually less than the initial cost of facilities and ongoing maintenance. Budgets need to realistically account for these "non-animal" expenses so that the project does not fail from lack of necessary initial support. For many projects the long-term payback ends up more than covering the ongoing expenses to support the project. Do anticipate, though, that the first few years will likely see a temporary cash-flow deficit.

IT'S MORE THAN THE HORSE!

Recent efforts have sparked great interest in the Choctaw horse for what many would never have considered a likely endeavor — competitive driving. Owners have campaigned several Choctaws with capable trainers, placing well in competitions such as the Ohio Combined Driving Event. Other successes by Choctaw ponies in eventing and endurance riding have helped to expand numbers of this critically rare strain of Colonial Spanish horses.

The initial cash outlay for animals with this potential for success can pale in comparison to the ongoing costs. Horses need training as well as tack and gear. For some endeavors, like driving, these can easily be more expensive than the original untrained horse.

Colonial Spanish Horse, Choctaw Strain

Choctaw tribal horsemen bred this strain for its highly valuable traits: comfortable gaits, endurance, and kind temperament.

Homeland: Oklahoma

Traditional uses: Riding, cattle work, driving

Size: 13.2–14.2 hands (54–58 inches)

Colors: Most colors known to horses

Other traits: About a third of Choctaws are gaited

Conservation status: Critical

Be Sure of Your Commitment

Each heritage breed has a history of productive partnership with its owners, and to continue that partnership requires committed people. Committed breeders are essential to ensuring that valuable animals, and the genetic contributions of those animals, are never lost. This does not mean that breeders can never quit — but it does mean that when the time comes to disband a herd or flock, it needs to be done carefully so that animals can go to other breeders to make a continuing contribution.

Keeping Track of Your Animals

NO MATTER WHAT SPECIES AND BREED you choose for your farm, if you plan to breed animals and are serious about contributing to the future of the breed, you should plan to use animal identification and record-keeping procedures. These range from simple to complicated, expensive to cheap, and each has a place in breed conservation endeavors. Choose your system wisely so that it serves you well for many years. Ask older and experienced breeders what they have used in the past, as well as picking the brains of younger and innovative breeders for ideas that might help the breed on into the future.

Animal Identification

Part of your commitment to a breed includes making sure that your animals are adequately identified. "Adequate identification" means specifically that someone other than the breeder can validate animal identities if ever needed. This can be done with a record book, an online record, or some other ready resource that can link individual animal identification to individual records of identity and production. *Identifiers* range from brands, ear tags, and microchips to complete photo records in breeds where most of the animals are uniquely marked or spotted. Birds can be identified by toe punches, wing bands, or leg bands.

Ear tags are generally effective identification for many species.

Most identifiers are attached or applied early in an animal's life. Unfortunately, nearly all identifiers will be in situations in which they can be lost. Number brands are likely the most permanent, but these are unacceptable to many breeds and breeders for a host of reasons. Tattoos can fade with age, microchips can migrate (and sometimes can be lost), and ear tags are notorious for being less permanent than desired. In well-identified herds and flocks, the loss of identification of a few animals rarely impedes their accurate identification, simply because their herdmates are all still identifiable and the process of elimination can correctly identify the one lacking an identifier. Nonetheless, re-identify the animal with a repeat tag or other identifier, or with a new one that is noted in the record. The safest approach is to always have every animal identified so that an interested outsider could match animals and pedigrees, even in the absence of the owner. This method greatly helps when an owner is incapacitated and unable to help the identification process along.

Record Keeping

Record keeping varies from situation to situation, but usually includes full date or at least year of birth, sire and dam, and specific identifying characteristics or identifiers that have been placed on the animal. Many people will add production characteristics such as dates and numbers of offspring, growth rate, health and veterinary data, or other details important to the breed and production system. Various computer-based record systems are available, and successful breeders use a wide variety of these, but something as simple as a record book for a physical reference resource works well, too.

Extensive and detailed record keeping is relatively routine for producers of larger livestock but is less common with smaller animals. Extensive individual records are especially rare among poultry breeders. If your farm is home to a *conservation flock*, however, identification using wing tags, toe punches, or leg bands should be considered. With minimal training, identifiers can be applied while chicks are in the brooder.

Beyond identifying individual animals and recording their lineage and life histories, good record keeping includes recording both individual and group matings. These particular records are an essential part of the breeder's tool kit for both conservation and production breeding.

Each part of an animal's record serves a purpose. Detailed, accurate records include pedigrees (the ancestry of the animal) but are much more than just the pedigree. Detailed records should enable the current or future owner of an animal to determine the degree of relatedness of that animal

TERMS TO KNOW

CONSERVATION BREEDER

An animal breeder actively working with a breed to maintain and develop good genetic structure to the herd or flock, with attention to production characteristics typical of the breed. Takes care to produce and provide sound, productive animals to other purebred breeders.

CONSERVATION HERD/FLOCK

A group of purebred animals bred with attention to the genetic structure of the group, as well as to production characteristics and breed type.

to any other in the herd or flock. The listing and rating of an animal's traits (such as rate of growth and finished weight in meat animals, and reproductive traits such as litter size and interval) are good tools for monitoring progress in breeding for production improvements, so that successful matings can be repeated and the offspring tracked through multiple generations.

Record keeping and animal identification are also vital tools when a genetic defect is detected in a breed. By examining the pedigrees of affected animals, you can often trace the defect back to a specific animal or animals. Other descendants can then be identified so that the defect can be carefully eliminated without damaging the overall genetic structure of the breed.

Poor Records Hurt Breeds

The sad truth is that every year some breeders experience a circumstance (such as a disability or death or a fire) that results in not being able to individually identify all the animals they so carefully bred and tended. Occasionally, all the individual details of animal identification were locked in the owner's head, so if tragedy occurs, that valuable information is lost. Despite all of that person's work, it becomes impossible to include their animals in the purebred, identified population. Nor can these animals be used in breeding programs of other breeders because their identity (and therefore ancestry) is no longer known. In a sense, the animals are lost to the breed, and the entire breed then suffers the loss of future genetic contributions that these animals could have made.

For so many compelling reasons, identification and records need to contain enough detail that any informed and interested outsider could decipher them. Keeping animal records up to date and in good order is essential to the ongoing contributions of a breeding program. It greatly benefits your breed's entire population to register breeding stock if there is an active studbook for them. Registration adds value to the animals and is the best way to ensure buyers that they are acquiring purebred stock.

Marketing and Processing

IF YOUR PROJECT INVOLVES PRODUCTION AND/OR BREEDING, you will need a means of marketing your animals and their products, whether through the local auction barn, a dealer, or by selling directly to consumers. If you choose to sell meat directly to consumers, you will need a state or federally inspected facility within a reasonable distance, except (in most cases) for poultry. Only inspected slaughterhouses can legally process your animals into meat for direct sale to consumers or restaurants.

From Field to Table

Processing is a crucial piece of the puzzle that is often overlooked by owners just starting out with meat animals. If you plan to produce meat only for your own family, you may process the animals on your farm (keep in mind that the bigger the animal, the bigger the job). If you intend to sell meat to others, you are usually legally obliged to go to an independent plant that does custom slaughter for small producers.

Unfortunately, in many areas of the country it may be impossible to find an appropriate processing plant near your operation, and that alone may make or break your plans. This obstacle is especially true with poultry, although on-farm processing is generally less strictly regulated with poultry than with larger livestock, and quite a bit easier to do yourself. In any case, be sure to familiarize yourself with local regulations and know all of the options that will help you achieve your project goals. Most state regulations can easily be found through county extension agents or with your state department of agriculture.

How Will You Promote Your Products?

Marketing of livestock and their products is a challenging endeavor with great potential rewards. Farmers choose a variety of ways to do this, and no one size fits all. Approaches vary from simple to complicated, and the payback usually is related to the amount of effort a farmer is willing to put into this part of the enterprise.

Given the range of options and the amount of work involved for each, one approach is to take slaughter animals to the local auction stockyard and sell them there. For some species, such as sheep and goats, this usually works reasonably well and provides a decent price. In the case of cattle, the markets can heavily penalize animals that are not the standard size or breeding that commonly go through these channels, and this unfortunately means that many heritage breed animals are likely to be heavily discounted. This general rule has several local exceptions, however, so be sure to ask around locally.

At the other extreme are producers who market eggs, dairy products, and meat directly to their customers through farmers' markets or through on-farm shops. These require more effort, record keeping, and in many cases involve regulations and inspections of facilities by health authorities. The rewards for this extra effort can be great, though, and some farmers carefully cultivate personal relationships with their customers as part of their enjoyment in participating in their local community. When marketing directly to consumers, stay within legal guidelines, which vary state to state. State

Guinea Hog

Docile in temperament, good foragers, gentle, good mothers — the Guinea Hog is a valuable and economical addition to a small-acreage farm.

Homeland: Southeast United States

Traditional uses: Local meat production, snake and pest control

Size: 150–300 pounds

Colors: Usually black; also, rarely, red, blue, or other colors

Conservation status: Critical

departments of agriculture are the usual source of information on regulations in your area. On the positive side, many farmers' markets are thrilled to have participation by farmers producing animal products as it can round out the offerings available to consumers.

Direct marketing can also involve selling elite and distinctive products to the restaurant trade. This must be done as a one-to-one connection with farmers and chefs. Each must understand the needs and limitations of the other in order for the relationship to work successfully. When successful, though, the return can be quite rewarding.

Niche Marketing Helps Heritage Breeds

Heritage breed projects can contribute greatly to breed survival and some do not even involve animals. For example, Connie Taylor has been key in advancing the survival of Navajo-Churro sheep through several of her activities. Connie does raise sheep, but she also buys and processes wool from other breeders. By pooling the fleeces of this distinctive breed she has enough raw material to process breed-specific yarns that are avidly sought by a wide range of craftspeople. This has created a brisk demand for the fleeces, which gives good returns to the breeders producing those fleeces.

HERITAGE PORK IS HOT WITH CHEFS

Chefs have discovered the unique flavor characteristics of pork from heritage breeds. Restaurants around the country are featuring pasture-raised and heritage-breed meats in dishes ranging from new takes on the traditional pork chop to exotic creations influenced by cuisines of many lands.

Chefs Sean Brock and Craig Deihl in Charleston, South Carolina, exemplify the innovative approaches chefs have taken with heritage pork, and their original dishes have won them national recognition. Deihl has taken a keen interest in charcuterie and has found that the lard from heritage pigs such as the Guinea Hog is perfect for European-style salamis and cured meats.

Heritage pig breeders who have been successful in establishing markets with chefs and retailers cannot keep up! Many of them purchase and raise weanling pigs from other breeders to meet the demand, so much so that weanling pigs of heritage breeds are a sought-after commodity in many areas and may be hard to come by. A niche exists in this area of production for breeders who want to sell young market stock to other farmers.

CHAPTER FIVE

Basic Breed Maintenance

Heritage breeds can offer as much to a small-scale, sustainably minded farmer as a farmer offers to them. In return for the farmer's stewardship of the breed and its genetic resources, the animals can pay their way by producing milk, meat, eggs, and fiber, supplying draft power, or winning in the show ring. The low-input systems of today's sustainable farms and ranches, so well-matched to the adaptations and advantages of heritage breeds, are an essential key to conserving these animals. Their future is secured through productive partnerships with farmers and ranchers.

Wyandotte chickens, for example, were once a mainstay of poultry production but became eclipsed when industrial systems took over the bulk of poultry farms. Fortunately, breeders like Frank Reese keep breeds like the Wyandotte productive, beautiful, and contributing members of modern farmyards.

Welcoming a heritage breed to your farm gives you a tangible link with our agricultural history, a way to begin regaining our farming ancestors' strong and intricate connections among producer, place, and animals. This may not be simple or easy, but it can be deeply satisfying to anyone who values the past and wishes to contribute to the future. For these breeds to continue making contributions, they have to be in good genetic and numerical shape, which is practical to ensure by paying attention to a few details.

How Heritage Breeds Are Maintained

Heritage breeds can be viable in the long term only if animals are being raised on many different farms, for different purposes, as long as those are consistent with the breed's past history. Owners who keep animals only

Wyandotte Chicken

Wyandotte chickens were once a mainstay of poultry production but became eclipsed when industrial systems took over the bulk of poultry farms. Since it forages well and its small comb size reduces frostbite threat, this dual-purpose heritage chicken is a good bet for free-range systems.

Homeland: United States

Traditional uses: Egg and meat production

Size: 6.5–8.5 pounds

Colors: Silver- or golden-laced, black, white, buff, partridge, silver penciled, Columbian, blue, and red-laced blue

Conservation status: Recovering

for exhibition or production make contributions that are just as essential as those who maintain breeding flocks and herds. No matter what type of project you choose, with care and planning you can make a valuable contribution to the conservation of heritage breeds on your farm.

The basic principles guiding selection and breeding of heritage breeds impact everyone who interacts with the breed, and so are important to all involved with heritage breeds, whether they are actually breeding animals or not. These principles are introduced below in the discussion of specific projects you can undertake with heritage breeds on your farm.

HISTORICAL SELECTION PROCEDURES

Traditional selection looks at production within the context of good adaptation to the environment and the production system in which the animals are raised. Most historical selection procedures came about from close observation of animals, their productive potential, and what sorts of animals could be relied on to have high levels of easy-care production.

Examples abound of the interaction of selection criteria and animal function. One example is the subtle point of selecting goats for strong, upright pastern conformation, which is related to longevity. Another is teat number (more than 10) in gilts (young female pigs), which assures that the desired large litters will be able to suckle. Choosing future herd sires from the offspring of older, productive females is a common tactic in several traditional systems because it assures that the dam was well adapted, productive, disease resistant, and of good temperament. Her sons should be able to spread their good qualities widely among their own daughters.

Breed Maintenance Basics: Selection

Selection is the powerful tool by which breeders maintain productive, genetically healthy livestock. Selection is the choosing of which animals will not reproduce, which ones will, and the extent to which they will. Selection shapes the genetics of the population, whether a single herd or the entire breed. It does this by making favorable genetic combinations more widespread and by penalizing less favorable combinations. Selection is essential for heritage breeds if they are to maintain their place in productive agriculture.

Production improvements are quickest when animal numbers are sufficiently large to allow for intense selection by producers of potential breeding stock. Making improvements is a numbers game, because the best few of thousands are likely to be better than the best few of tens. Nonetheless, talented and observant animal breeders have always been able to produce superior stock from small populations, so numbers are not the whole game.

Important to keep in mind is that quick improvements (changes) are not without risk. Permanent genetic changes can result and must be carefully considered.

Especially in the case of landraces, selecting for production levels is more than just a numbers game because most landrace breeds have traditionally seen very little selection for production. In most landraces, selection more heavily emphasized adaptation and survival, with less extreme selection for production levels. Few owners of landraces have historically put a high premium on production, so that even the highest-producing animals could see little reproductive advantage over their more average herdmates. Many of the traditional systems typical of landraces lacked animal identification and single-sire matings that would allow for intense selection. Especially in extensive ranges, even catching all the animals could be an unrealistic challenge, let alone culling all of the substandard ones.

Because of the relative lack of intense selection in landraces, dramatic gains in production levels can be made in just a few generations. The high levels of genetic variation in landrace breeds give producers more to choose from and are wonderful raw material for selecting for production goals. Not only can production be enhanced but frequently these breeds bring with them their exquisite adaptation and disease resistance. The end result of a good breeding program is animals that are productive as well as adapted.

Case in Point: Spanish Goats

Spanish Goats have long served their owners in the Southwest by providing brush control as well as tasty meat. The original size was quite small, but breeders have been able to increase this by selection. The process took only a decade or so, and it changed the does from 60-pound weaklings to 125-pound productive powerhouses.

See pages 49 and 183 for Breed Snapshots of the Spanish Goat.

Balanced Selection Is the Key

The ideal breeding program for a heritage breed selects animals for *both* their ability to produce *and* the array of traits related to mothering ability, disease resistance, temperament, grazing or foraging behavior, and adaptation to the environment. If animal numbers are critically low in a breed, however, it is more important to rescue the breed first, and worry about an ideal selection program later.

Thus the primary focus of breeding programs should be responsive to changes in the current state of the breed rather than a static "one size fits all" approach. At one end of the continuum are breeds or bloodlines that need rescue because they have tiny populations, so the focus must be to conserve their genetics. At the other end are large populations that should undergo more intense selection for production, after their unique genetics have been maintained and other traits have been stabilized. Along the line between these two ends are several points where the balance shifts between selection for production, selection for other traits, and selection for genetic health of the population. Each point is useful for some specific and important heritage breed conservation projects.

HOW BREED POPULATIONS DETERMINE BREEDER PRIORITIES

LOW BREED NUMBERS	TRAIT	HIGH BREED NUMBERS
Low	Selection for productivity	High
High	Attention to bloodlines	Low
High	Selection for genetic health of breed	Low

BREEDING TERMS TO KNOW

BREEDING. Selecting animals and then mating them to produce the next generation

CROSSBREEDING. Mating animals of two different breeds

CULLING. Removing animals from the reproducing population. This may involve slaughter for many species, but for horses can include gelding extra males, or using steers for oxen.

INBREEDING. Mating animals that are closely related, usually first-degree relatives such as parent/offspring or brother/sister

LINEBREEDING. Mating animals that are related, but not as closely as first-degree relatives

MATING. Pairing animals for reproduction

OUTBREEDING. Mating animals that are not related — a general catch-all term that covers both crossbreeding and outcrossing

OUTCROSSING. Mating animals that are not related and that are from two different bloodlines within a single breed

SELECTION. The process of choosing some animals for reproduction, and removing (culling) other animals from the reproducing population

The issues of selection can become complicated and even overwhelming, but an orderly approach can help to simplify them so that they make as much sense in the farmyard as they do on paper. Sorting through your animals to note the ones that produce best in your management system and environment is an important task. In general, by matching the weaker ones with stronger ones you will help raise the level across all your animals, and also for the whole breed.

Diverse Bloodlines Serve Breeds Well

Despite their name, Java chickens are an American breed. They are also a very old breed and have been a feature of U.S. farmyards for many generations. This breed suffered especially severe population losses in the face of competition from industrial birds, so much so that a once-common breed became very rare. Fortunately, a handful of breeders kept large flocks of these distinctive birds going through the lean years.

See page 132 for a Breed Snapshot of the Java chicken.

Discovering and validating these flocks is an essential part of heritage breed conservation. Dedicated enthusiasts, old-timers and newcomers alike, have discovered some critically important portions of a number of breeds that would otherwise have become extinct.

This work is never finished. The Livestock Conservancy first became aware of Garfield Farm Park's line of Javas, along with those of Duane Urch and the Bowens. These were the start of a plan to reinvigorate the breed back to its historic productive potential. Later finds were the Jansen flock in the hands of a dedicated elderly breeder whose family had the birds back in the 1940s as part of a hatchery stock that the family used for sales. These, along with the recently encountered Ward flock, were able to contribute to the breeds' currently enhanced prospects. In 2012 yet another line, Wear, surfaced and can now contribute to the conservation of this old heritage breed.

Numbers Can Make the Difference

Sometimes whether these principles are applied, and to what degree, are dictated by numbers. They may be delayed completely for the rarest of the rare (whether breeds or bloodlines), for which the focus is on expanding numbers without any risk of losing genetic material beyond what has already been lost. In contrast, more numerically robust breeds can readily afford to lose animals to culling without endangering the genetic structure of the breed, and in those situations good basic breeding principles should be applied. The breeds with higher numbers therefore have more opportunity for breeders to select animals for enhanced production.

Java Chicken

Javas are especially valued as docile birds, good foragers, and gentle, good mothers.

Homeland: United States

Traditional uses: Meat and egg production

Size: 6.5–9.5 pounds

Colors: Black, white, mottled, auburn

Conservation status: Threatened

With the especially rare heritage breeds, consider both their genetic structure and the need for maintaining and improving production levels. Ignoring either of these results in a diminished future for the breed.

Select across Multiple Families for Long-Term Success

Landraces such as the Myotonic (Tennessee Fainting) goat have historically been subjected to very little selection, especially for production traits such as growth rate and conformation. Most past owners simply used them for their own consumption, so growing these for a commercial market was unimportant. That has changed with the recent upsurge in demand for goat meat. Selection in the early days was geared more toward survival and adaptation, and was imposed more by nature than by the owners.

This lack of intense selection provides today's breeders an ideal gene pool for making rapid progress in growth rates and meat conformation. The breed has benefited from production selection in recent years, and its numbers have grown significantly. Their conservation status has changed from Threatened to Recovering, a true success story.

See page 96 for a Breed Snapshot of the Myotonic goat.

Torsten and Phil Sponenberg have raised Myotonic goats since becoming alerted to their precarious situation in the early 1990s. They use the farm name Beechkeld in tribute to the trees and springs on the farm, located in Virginia. Each year Beechkeld uses six male goats to produce kids from groups of females that number from four to twenty. Each of these buck goats is used for only one year, and then he is sold either to other breeders or at the local livestock market. One male kid from the half-brothers from each sire is then chosen for breeding the next generation to ensure that the male lines continue in the herd. The does get reshuffled from group to group annually, based on the male being used, his pedigree, and his other characteristics.

In the early years of this practice the temptation was to select all of the buck kids from only one or two of the groups of half brothers, specifically the ones with buck kids that were noticeably bigger, smoother, and faster-growing than the sons of the other males. Any insistence on retaining males from the somewhat less ideal families seemed counterproductive. Following the multiple-line approach rigidly year after year, however, has yielded long-term benefits, such as the broad genetic base for the herd that came from using multiple families. In addition, the ongoing selection from year to year has raised the overall quality of even the lines that were originally weaker.

Over time, selecting the best males from each family rather than from only the single best one has resulted in an increasing number of families producing males of similarly high quality, while also maintaining diverse genetic lines in the herd. The process took a decade but secured the genetics of the herd for the future. This, combined with broad selection for females to add back into the herd, has produced a high quality as well as broad genetic base.

The short-term fix of selecting males from only the single superior family in the early years would have resulted in excellent quality to start, only to decline in later years as the herd genetic structure would have closed in on itself. Inbreeding would have become unavoidable and with it, the negative consequences.

Special Issues in Maintaining Heritage Breeds

As the art of maintaining heritage breeds evolves, a few issues have come to light that are important for new breed stewards to know about. Each breed comes with its own specific issues, including strengths and weaknesses both in the animals themselves and in the breeder community that stewards them. Success is a tricky balancing act of animal and organizational skill, and any of several snags can sink an otherwise successful program.

The Dangers of Playing Favorites with Breeds

One important issue is the trap of thinking that any single heritage breed is the magic silver bullet for all owners and all situations. This trap is especially true with poultry, for which "end use" producers (eggs and meat) must avoid believing that a single heritage breed is the only decent alternative to the birds used in industrial confinement systems.

This problem has occurred to some extent with Buckeye chickens as well as with Bourbon Red turkeys. Both of these breeds have deservedly gained a reputation for excellent production and adaptation to sustainable farming, to such an extent that in some areas other rare breeds are ignored, even though they may be an equal or better fit for many producers. The result is the favoring of one heritage breed above all others, and thus the overall issue of breed rarity remains much as it was except for the single now-popular heritage breed.

The Bourbon Red turkey has become a good strong candidate for heritage turkey production. It has benefited numerically from more and more producers choosing this breed over the industrially produced Broad Breasted white birds. Other types of heritage turkeys can also benefit from this boost of demand, but only if the Bourbon Red does not become "the option" instead of "an option" for producers interested in alternatives to the industrial birds.

In many cases, only luck or an immediate opportunity has caused a specific breed or variety to be chosen for a project or recovery program by

See page 87 for a Breed Snapshot of the Bourbon Red turkey.

The Livestock Conservancy and its members, and the choice had little to do with inherent superiority for production. Producers need to look broadly at what is available, and to choose a turkey, broiler, or egg-producing hen from a breed that is up to the task and that also needs a boost. Several breeds may be equal to the challenge of providing production along with presenting important conservation opportunities for the people that work with them.

Each breed embodies its own combination of traits that makes it a good choice for specific production systems, and matching the breed to the system is part of the appeal of heritage breeds. As soon as a single breed is touted as best for all situations, the close pairing of breed with location disappears.

The Dangers of Playing Favorites with Bloodlines

As the movement to conserve and use heritage breeds has gained steam, it has spawned a demand for animals that are true to their historic roots and productive as well. The Livestock Conservancy has proudly and ably led this movement, encouraging broad participation by lots of breeders using many breeds and bloodlines.

While breed purity is important, and bloodlines can differ in their productive potential, The Livestock Conservancy has never designated some bloodlines as "Livestock Conservancy recognized" while not endorsing others. Vitally important is for *all* purebred bloodlines of any breed to go on into the future. While some may have specific strengths, breeds need a wide variety of bloodlines to continue on as viable and healthy genetic resources. Once breeders discard some bloodlines in favor of a few select ones, the whole issue of resource integrity rears its head and endangers the breed. Variety is the spice of life *and* the key to breed survival.

Slipping Production Due to Low Breed Populations

A second concern for all stewards of heritage breeds: when your chosen breed has an overall low population, it is difficult to effectively select for production. Slipping production dooms any breed to a bleak future, so it is essential for all breeders to select for animals who produce well. Excellent animals can and do come from even the smallest herds and flocks, as long as the breeder is paying attention and selecting for only the best in the group.

In the smallest populations it can be important to focus on expansion of numbers, ensuring a good balance of genetic representation of the animals available. After numbers increase, focus more attention on selecting for production.

Popular-Sire Syndrome

In many breeds, having a "popular sire" arise is all too common. This happens in all species, and can be through show-ring wins, promotion by a successful breeder, or even by having proven through genetic testing that a particular sire tops the competition for a specific trait. When the sons of a popular sire are used widely throughout the breed, after a few generations this one family has diminished nearly all others. This strategy is not always bad if production potential of this one family is indeed that much better than all others, but if a one-sire family *replaces* all other families or strains within the breed, it can lead to difficulty in managing the breed's genetic structure over the long term. Such a situation quickly leads to inbreeding, and inbreeding always threatens a decline in vitality and health.

The solution is that unrelated males should (ideally) always be available to breeders. Making this a reality in the long run is only possible if breeders are paying close attention to their own breeding programs *and* paying close attention to the breeding programs of others. As long as unrelated males are available, all breeders will be able to find a suitable outbred mating for every female. This approach is insurance against future inbreeding depression. To make sure enough unrelated males are obtainable for long-term breed health, breeders need to avoid the "popular-sire" syndrome.

See page 138 for a Breed Snapshot of the Cleveland Bay horse.

Cleveland Bay Horse

A favorite of Buffalo Bill Cody and both Queen Elizabeth I and II, this ancient, critically rare breed excels in athletic ability, whether for riding or driving.

The rarity of Cleveland Bay horses is recent, with numbers falling over the past several decades. In addition, sire lines (families descended from individual stallions) have also diminished, so that fewer and fewer options are available to breeders. The result is that breeders have difficulty in locating stallions that are unrelated to their mares.

Homeland: Cleveland, England

Traditional uses: Riding, driving

Size: 1,300 pounds

Special adaptation: Athletic ability

Color: Bay

Conservation status: Critical

Selection for Production Only

Selection for production is important, but if production level is the only strategy for identifying and retaining breeding animals then the population can hit various snags. The major snag can be inbreeding and the resulting loss of overall vigor. A second snag can be short-sighted focus on only certain aspects of production such that overall balance in the animals becomes lacking. This misstep is especially likely with high levels of show-ring competition where extremes are selected, rather than overall balance.

Selection based solely on production has been rarely true of heritage breeds in the past. Such an approach is more common in modern production breeds, and especially in industrial livestock. In these breeds the genetic structure can become constricted around the contributions of a few excellent producers and animals, and even in numerically strong breeds this can cause problems from inbreeding.

A "PRODUCTION-ONLY" FOCUS CAN BACKFIRE
A focus on production alone, especially in a carefully controlled environment, results in exquisitely productive animals. These animals, though, have generally lost a great deal of their other adaptations, and so cannot be expected to maintain their high levels of production outside of that narrow environment in which they were selected to produce.

CORNISH-ROCK BROILER, INDUSTRIAL CROSSES

Specially designed for industrial systems, the Cornish-Rock broiler is adapted for high feed efficiency and rapid growth.

HOMELAND: United States

TRADITIONAL USES: Industrial broiler production

SIZE: 5 pounds at 40 days

COLOR: White

CONSERVATION STATUS: Not ranked; population in the millions

CHAPTER SIX

Choosing a Heritage Breed Project for Your Farm

ERITAGE BREED PROJECTS VARY considerably in both purpose and scope, and each contributes to the future of any breed. No single project is best for all owners and all situations, and having lots of owners doing things differently from one another means the genetic material of the breed will maintain diversity and vigor. A successful and rewarding project for you depends on a firm foundation of thoughtful planning, and above all matching the right project to the right person, the right facilities, and the right place.

Projects are bounded only by the interests, abilities, and imagination of those undertaking them. A key point is to focus on a single project and to avoid drifting from one to another. A targeted strategy usually leads to success, but untargeted projects do not last long, rarely succeed, and can do long-lasting damage to the breeds involved. The damage comes from the projects using up genetic material and then not putting it in good shape for the next generations of animals, and also for the next generations of breeders.

Two basic categories of projects include 1) those that involve nonbreeding populations and 2) those that involve breeding animals. The first category includes owning animals for show or exhibition, or raising purchased young animals strictly for end-use production. The second category includes breeding your own animals for both production and conservation, and breeding animals strictly for breed conservation or rescue. All breeding projects require a sound understanding of traditional selection methods.

Projects with Nonbreeding Populations

IT MAY SEEM CRAZY TO DISCUSS nonbreeding populations in a book about heritage breeds and their survival, because no breed can survive without a breeding population. In many species, however, the nonbreeding animals make essential contributions to the overall health and security of the entire population. For some species, a secure future of any sort is difficult to imagine without these types of projects.

Project One: Performance and Exhibition

Performance and exhibition animals can be found in nearly every species but are most commonly associated with horses. Most demand for horses is based on athletic ability of one sort or another, and not on human consumption. This selection criterion is different from other species, although a few breeds, such as Milking Devons, include within their demand structure a brisk trade in steer calves for development and training as oxen.

For example, a huge segment of the horse population is not involved in reproduction in any way. The horse industry is now dominated by performance horses, and heritage breed performance horses that are good at what they do contribute greatly to the demand for that breed. Such horses and their owners are great ambassadors to potential new owners.

Project Two: Production Only

Production of milk, meat, eggs, and wool is a logical fit for many heritage breeds, and such animal projects most readily come to mind when "farming" is mentioned. These projects have broad appeal and involve a large number of heritage breeds. This type of project is based only on production of the end-product, so that reproduction of the animals is not required. These

SHORT-TERM PROJECTS CAN BE WORTHWHILE

Not all contributions to livestock conservation need to be made over long-term breeding careers. A great many significant contributions can be made over a short time by any committed person with a careful plan and excellent follow-through. The most dramatic of these is rescue of a breed or of a bloodline within a breed. Committed breeders can make all the difference in the survival of these rarities, and can build them up to health and good numbers in just a few years in order to make them available to other breeders.

projects have broad appeal for farmers not having the facilities or the interest in breeding animals.

Restoring heritage breeds of poultry to their previous levels of production is essential to the expansion and promotion of these breeds. Targeted selection greatly improves production in only a very few generations if enough offspring are produced to provide a wide choice. Success by The Livestock Conservancy with the Buckeye breed for meat and egg production is now being expanded to other breeds, such as Javas.

The Livestock Conservancy's work with the Buckeye breed involved hatching out hundreds of chicks and working with cooperating producers to evaluate these as breeding stock. Attention was paid to growth, vigor, meat conformation, and egg-laying potential. Within a few years the project had produced birds that were vigorous, productive, and fit the breed standard so well they were winning at the shows. The result is a handsome, productive, and healthy bird well suited for farmyard production. High-quality, productive birds ensure demand for a breed. With work and time, they have the potential to return profit to their owners as well as enjoyment.

Short-term contributions can include anything that builds demand for purebred animals. In some cases, this may be animals specifically raised for meat production, as long as the breed identity stays with the products and boosts demand for the breed. Seasonal production of poultry is one short-term project (generally just a few months each year) for many producers and

SELECTING FOR PERSONALITY

The Buckeye has the reputation of being a very laid-back, easygoing breed of chicken. But because their originator, Ms. Nettie Metcalf, of Warren, Ohio, incorporated Black Breasted Red Game chickens into her breeding scheme, the Game's aggressive nature can sometimes come to the surface. One strain, in fact, was aptly referred to as the "mean-as-snakes" flock. Fortunately, over the years most breeders have selected for birds with a mild temperament to keep this large breed clear of aggressive tendencies. The take-home lesson is this: when searching for animals, *always* be sure to ask what selection criteria are used on the source farm, including personality and temperament.

TERMS TO KNOW

SEEDSTOCK

Animals specifically intended for reproduction. Not all purebred animals are destined to contribute to the next generation of the breed, and "seedstock" designates those that are.

SEEDSTOCK PRODUCERS

Breeders who target the sale of purebred stock for breeding purposes. In mainstream modern breeds, only a minority of breeders are seedstock producers. In heritage breeds, most breeders should be seedstock producers, even though that takes additional effort in selection, identification, and promotion. A breed with a few seedstock producers rests on a narrow genetic base influenced by only a few selection programs. A breed with many seedstock producers rests on a wider, and stronger, genetic base.

is a contribution that has been essential in raising demand for turkey poults and the young of other species as well.

This type of contribution is by no means limited to poultry, as a brisk demand for heritage breed products is essential to conserving breeds of all species. For example, a producer buying heritage breed weanling steers to fatten for market is greatly helping the *seedstock producers*. They provide demand for male calves (and heifers) that are not needed for conservation breeding. In some areas of the country the demand for heritage pork outstrips the supply, so that producers who buy weanling pigs and finish them out can reap good economic rewards for their effort while at the same time benefiting heritage breed survival.

Brisk demand for a breed based on rock-solid production creates a positive feedback loop. Increased demand fosters larger population size; a larger population size creates more choices for breeders to select breeding animals in order to enhance production traits. Enhanced production traits generate even more demand for the breed and its products. Everyone wins.

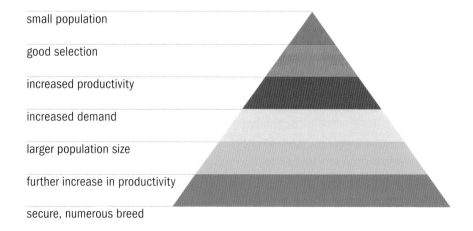

small population

good selection

increased productivity

increased demand

larger population size

further increase in productivity

secure, numerous breed

The Fine Art of Culling

Culling is something of a balancing act. Too little culling results in weak and substandard animals. Too much culling can result in a gene pool that is too small to function well. Unfocused culling can result in a genetic structure that is awkward to maintain long term. Sometimes not enough superior animals are on hand to choose from, and some animals with lower production potential must be retained simply to provide the genetic variation they can contribute.

Yet animals with mild to moderate defects in performance, conformation, or breed standard can be used effectively by their owners to contribute to breed security. These *second tier* animals can make that contribution only if they are mated wisely and if the resulting progeny are used appropriately. Also essential is that breeders recognize that these animals are indeed second tier. They should never be confused with excellent-quality, elite breeding animals. Their use should be short-term, strategic, and for specific goals.

The history of every heritage breed (except ferals) has involved breeders actively culling their stock. Careful selection shaped their animals, fine-tuning both adaptation and production. Without continuing selection (a nicer way of saying "culling"), key components of any breed's traditional character will be lost. Breeders who want to keep every animal in their herd or flock must absolutely ensure that only selected animals are bred and must have the resources to support an ever-expanding herd size.

Many factors influence selection.

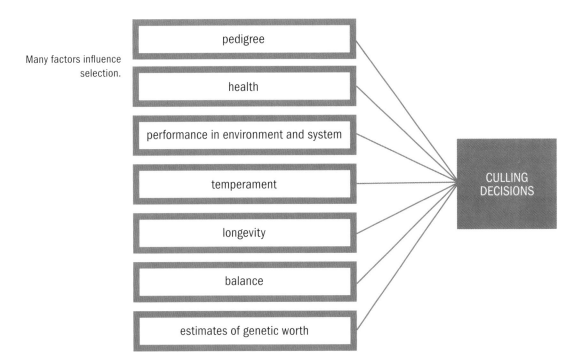

Selection Strategies in Pigs

Hog breeders have an advantage over sheep, goat, cattle, and horse breeders: large litters of animals, usually two litters per year, are produced. This provides good numbers from which to choose replacement stock. Those numbers help, though, only if the breeders actually make the selection decisions. Ignoring selection for production characteristics in breeds like the Red Wattle, where production was traditionally selected for, is a disservice to the breed's recognized ability to produce high-quality pork.

Josh Wendland is a lifelong farmer in Kansas who maintains breeding herds of Red Wattle hogs, along with Mulefoot and Gloucestershire Old Spots hogs. He does it right. He pays careful attention to productivity and fertility in his hogs, saying that "if a sow cannot produce and raise successfully at least six piglets per litter, then she does not belong in the breeding program."

See page 148 for a Breed Snapshot of the Red Wattle hog.

Wendland also pays close attention to behavior. While some breeders often select boars primarily for their conformation, Josh also emphasizes selection for how his boars behave toward females. Boars that are not vigorously interested in sows at all times are not likely to be chosen as breeders.

Josh Wendland's attention to selection details has kept his Red Wattle hogs attractive and productive.

Red Wattle Hog

This breed is highly valued for its hardiness, foraging ability, rapid growth, and gentle temperament.

Homeland: Texas

Traditional use: Pork production

Size: Usually 600–800 pounds, but some boars weigh up to 1,200 pounds

Colors: Red, varying from light to nearly black

Other traits: Paired wattles on the neck

Conservation status: Critical

Project Four: Rescue for Rare Breeds or Bloodlines

Most heritage breeds are small in numbers and have been that way for a long time. In some cases, numbers are so low that the breed is in imminent danger of extinction. Therefore, every mating decision needs intense attention in order to grow breed numbers and to retain all the genetic material in those few animals. The situation with individual bloodlines within many breeds is the same, and these bloodlines also need the same sort of "rescue first aid."

Rescues are tricky and complicated. Detailed record keeping is essential, because accurate pedigrees are the only tools available to make certain that matings will result in a completely balanced representation of the complete genetic material. In rescue situations, selection for other criteria such as production takes a back seat to the urgent short-term need to save the breed. Breeding decisions in rescues usually need to be made for genetic considerations and nothing else, in an effort to conserve the overall genetic structure of the breed or bloodline. The goal is to ensure that the more genetically unique animals, each of which contributes to the breed's genetic structure, are represented as broadly as possible in successive generations. Survival and expansion of the genetics leaves the breed in good shape for selection in the future when the breed population is larger.

A rescue project can take several forms: rescuing a rare bloodline or breed, rescuing a breed by combining diminished strains, and rescuing a single animal. The specific breeding strategies used differ for each case and are also determined by the details of the situation. The technical details of this important topic are discussed in chapter 8.

Rescues Save Breeds from Extinction

Santa Cruz horses hail from an island off the coast of California. Ranching ceased on the island in the early 1980s, leaving only a few horses remaining from the herds of local island horses that had survived there for decades. These ran free, reproducing a few individuals until 1997 when the horses were slated for removal by the National Park Service. About 15 horses were rescued by committed breeders, and most of the horses were related to one another. The rescuers needed to give careful consideration to each animal's role in securing the future of the breed. Fortunately, DNA work was able to establish relationships among the horses.

Two of those committed breeders are Christina Nooner and her husband, Troy. Their Sunshine Sanctuary for Kids and Horses was founded to provide a therapeutic environment for children in crisis. Conserving the Santa Cruz horse breed was a logical fit. It all began when a sick and malnourished abandoned Santa Cruz filly named Sunshine arrived at the

Nooner ranch. Christina found immediate veterinary care for the animal and used her nursing skills to bring the foal back to health.

Early on, Christina realized that this patient and gentle horse was not only special but also characteristic of the breed. The naturally calm demeanor of Santa Cruz horses made them perfectly suited for the children's hippotherapy programs at the Sunshine Sanctuary. With that in mind, she decided to become involved with the conservation of these horses on the ranch.

Through the on-site programs and the sanctuary's promotion of the breed by taking horses to local horse events, Santa Cruz horses are beginning to be recognized in their native California. Christina's careful breed planning, with guidance from The Livestock Conservancy, has resulted in a slowly increasing population and new breeders in several locations. Still, the horses are at great risk with fewer than 50 individuals.

Early and effective action in this case has had good success in saving the unique genetic heritage of these horses. Future generations can now benefit from the availability of their unique talents.

Christina Nooner's attention and care rescued the Santa Cruz horse from extinction.

Colonial Spanish Horse, Santa Cruz strain

This gaited, island-born breed is esteemed for its good temperament and smooth gaits.

Homeland: Santa Cruz Island, California

Traditional use: Riding

Size: 14–14.2 hands (56–58 inches)

Colors: Chestnut, bay, buckskin, palomino, or cream

Conservation status: Critical

Generations of Commitment Save a Breed

Few people make a commitment to the extent Bryant Rickman has with his Colonial Spanish horses. The Rickman herd is made up of bloodlines developed by several Native American nations — Choctaw, Cherokee, Huasteca, Kiowa, Comanche, and others — along with Colonial Spanish horses from Utah and New Mexico. A famous mustanger, Gilbert Jones, began collecting the breed in the early 20th century after he noticed that the Spanish mustangs once common among Native Americans in the West were quickly being replaced by larger breeds of horses. The larger breeds were showier and more impressive but lacked the endurance of the original, smaller Colonial Spanish horses.

Beginning with the purchase of his first Colonial Spanish stallion, an iron-gray horse he named Grey Eagle, Jones spent many years searching for and buying exceptional horses, including many from the herd that produced the famous endurance horse Hidalgo. Gilbert continued to breed and conserve Spanish horses until his death in 2000, when his herd passed on to his good friends Bryant and Darlene Rickman. Today the Rickmans continue the work Gilbert began so many years ago.

Until 2007, the Rickmans' horses were largely kept on open range in the Kiamichi Mountains of southeastern Oklahoma near Blackjack Mountain and Medicine Springs. In 2007 the grazing lease on this privately held timber land was terminated and the horses had to be removed. Using catch pens situated in various sites in the mountains, Bryant captured 300 horses within the herd with the help of friends and family and then relocated them to his ranch in Soper, Oklahoma. Because the ranch could not support the entire herd for long, with the guidance of The Livestock Conservancy, new owners and breeders were found for a majority of the herd. Today, with the help of this new generation of owners and breeders, these rare Native American strains of Colonial Spanish horses have been secured.

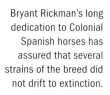

Bryant Rickman's long dedication to Colonial Spanish horses has assured that several strains of the breed did not drift to extinction.

Colonial Spanish Horse

These historic horses have good temperaments, are easy keepers, and have tough, durable feet. Many Colonial Spanish horses are *gaited*, meaning they have a smooth four-beat gait instead of the jarring two-beat trot of most horses.

Homeland: Southern and western United States

Traditional uses: Ranch work and endurance riding

Size: 13 to 15 hands; 600 to 750 pounds

Colors: A wide variety — most colors and patterns known to horses

Conservation status: Endangered

Matching Basic Motivations with Specific Heritage Breed Projects

UNDERSTANDING WHICH OF THE MOTIVATIONS outlined in chapter 1 (conservation, cultural connection, and production of commercial commodities) matter most to you can help you determine which projects are likely to be most rewarding. None of these motivations is at odds with major types of projects, but some of the endeavors discussed above have very special needs that are best met by breeders and owners with specific goals. Matching these up is not absolute but does help in some cases to make that final good fit of person, animal, and project that is so useful for long-term success.

Before too many years have passed, you, too, will have gained valuable experience and will be a good candidate for even the most challenging projects.

PRIMARY MOTIVATIONS FOR HERITAGE BREED PROJECTS

Commercial Products

Conservation

Cultural Connection

PROJECT: **Non-Breeding**

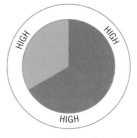

PROJECT: **Selection/ Improvement**

PROJECT: **Rescue**

PROJECT: **Conservation and Production**

PROJECT: **Production Only**

CHAPTER SEVEN

The Art and Science of Breeding

BREEDING ANIMALS IS BOTH AN ART and a science. Some breeders prefer the scientific side of it, while others prefer the art of shaping their herds and flocks into their mental picture of perfection. Both approaches have merit, and strong programs usually rely on both rather than entirely on one or the other.

This chapter outlines some aspects of these approaches that can help the breeder to successfully plan and accomplish a heritage breed project.

Basic Principles

BREEDING ANIMALS IS THE PROCESS of selecting and pairing animals for reproduction with the aim of producing the next generation. A fascinating and challenging endeavor, animal breeding involves general management as well as insight into how genetics works within populations. Success is the goal, both for the individual herd or flock, as well as for the breed as a whole. Different strategies are involved in different situations, but all of these should work together to result in productive, healthy stock that are a delight to their owners, all the while producing top-quality meat, milk, eggs, wool, and other goods and services.

Traits and Their Heritability

Traits simply means characteristics, and each characteristic is the result of an interaction of genes with the environment. Some traits are not influenced significantly by the environment and therefore are controlled mostly by genetics. For example, black cows are usually pretty black regardless of where they are, so the trait *black* doesn't vary much because the environment's influence is minimal. Other traits have lots of environmental influence. For example, a genetically high-producing Holstein in a swamp in Venezuela is unlikely to produce much milk because the environmental conditions can veto even the best genetic material's ability to express itself.

The relative contributions of environment and genetics lead to a number called *heritability*, and while the details can become complicated, the simple picture is that most traits can be considered to have low, medium, or high heritability. Selection is more likely to rapidly change those traits with high heritability, because they have more genetic influence. A complete list of traits is impractical, but for many breeds and species the following are important, with an indication of their relative heritability:

MEAT TRAITS — MEDIUM TO HIGH HERITABILITY
- weaning weight
- conformation score
- average daily gain post-weaning

MILK TRAITS — LOW TO MEDIUM HERITABILITY
- milk yield
- fat percentage
- protein percentage

FERTILITY TRAITS — LOW HERITABILITY
- age of puberty
- interval between births
- ovulation rate/litter size

HEALTH TRAITS — LOW HERITABILITY
- longevity
- disease resistance

Although it is tempting to believe that the breeder cannot do much about traits of low heritability, it turns out that in many cases selection is indeed effective in improving these traits. Some, such as health traits, are extremely important and need constant attention so that animals are viable and productive.

Large Black Pig

Conformation, growth rate, and meat characteristics are all highly heritable in pigs. Reproductive rate is less so but equally important to the final success of breeds such as the Large Black pig. Good foragers and excellent mothers, Large Blacks have found a strong market among today's chefs.

Homeland: England

Color: Black

Traditional use: Pork production on range

Conservation status: Critical

Size: 600–800 pounds

General Principles for Beginners

SELECTING AND BREEDING ANIMALS is a worthwhile endeavor, and good breeders spend a lifetime learning just what works best. They do this by closely observing what happens with their own animals and also by seeking out information and approaches from others. The careful and successful animal breeder is likely to change and modify approaches as experience is gained, but that does not mean that starting out is hopeless! A few general principles help folks begin who have little or no experience.

BREED FOR BALANCE. Animals are a combination of traits: some might be strong, and some might be weaker. The goal should be to correct the weaknesses with strengths in the mate. For example, a ewe that has good meat conformation, wool quality, and temperament but is second tier in milk production might be matched with a ram whose daughters have proven to be "milky." Her offspring will benefit from inheriting the good characteristics of the dam while improving in the one area she is weak through the contributions of the sire.

CULL ANIMALS THAT ARE DIFFICULT TO WORK WITH. Temperament is important! Keep in mind that rogue animals who are aggressive, flighty, or prone to panic easily are a risk to breeder and breed alike, regardless of their production potential. Rogue animals in most cases should have no role in the breed's future.

SELECTING FOR MINIATURIZATION AND OTHER ODDBALL TRAITS

Selecting for extremes can lead breeders and their animals down blind alleys. In most cases, balance is the end goal so that animals are functional and healthy. While dog breeds are notorious for extremes, the same can also be true of livestock. Some traits, like miniaturization, can bring with them unforeseen consequences as sub-vital dwarves are produced.

In Dexter cattle, for example, at least two types of genetic mechanisms lead to the resulting small size. One of these (the short-legged type) leads to the stillbirth of defective "bulldog" calves when they have inherited a double dose of the causative gene. The other (long-legged) type does not encounter this problem. The tip-off is that the long-legged type is much less extreme, and avoiding extremes would have avoided the problem of those bulldog dwarf calves.

In general, avoid selecting for radical extremes, because these frequently bring with them deleterious health-related traits. Selecting for big, rangy Suffolk sheep, for example, is the likely culprit in the high incidence of "spider lamb" syndrome in that breed. The spider lambs are long, gangly, and rarely do well.

USE A BROAD VARIETY OF THE BREED'S GENETIC RESOURCES. Herd sires can be a major constriction point for many herds. Important to keep in mind, though, is that most of the genetic variation in a herd is in the females, not in the single or few males that are used. One strategy to maintain that variation is for breeders to make sure that candidate sires come from all portions of the herd.

Scientific Selection Methods

Breeders of many breeds of livestock now have even more tools in their toolbox. These tools include a host of varying approaches, some called Estimated Breeding Value (EBV), or Estimated Progeny Difference (EPD). The key to understanding these is the "E" for "estimated." Each of these depends on measuring the performance of the animal and also his or her relatives. The higher the number of animals that are measured, the greater the reliability of the estimate.

These estimates can be quite useful in making breeding decisions. For example, an EPD for an increase in yearling weight might be very welcome in a breed used for meat production. An estimate for lower birth weight, in contrast, might be a good choice for a male to mate to first-time females so that the births are easier on everyone.

The catch for these techniques and estimates is that they are unlikely to be available for many of the heritage breeds, because the quantity of animals and the record keeping involved in crunching the numbers are simply not available. So, while these can be useful techniques, they may not be practical in all settings. They are nearly always limited to situations in which owners and breeders are heavily promoting and investing in individual animals.

A more recent development is *genomic selection*, in which genetic variants are measured right in the DNA of the animals. This technique has great potential for guiding some decisions, because certain combinations of DNA are especially valuable in raising production levels. Those components might actually be present in moderately productive animals, and if they are paired up with the right mate, the offspring can have the combination that comes up a winner.

An important background assumption on genomic selection is that it works only in situations for which the various combinations have been investigated as to their final result in whatever production character is under consideration (for example, growth rate or milk production). Many heritage breeds might well have alternative genetic mechanisms for these traits when compared with more modern breeds, and a concern is that these genetic mechanisms may go unrecognized when using genomic selection techniques.

Livestock Selection Decision Making

Selection decisions need to be made on a host of factors, which can pull in a variety of directions. Having some sort of organized flow to these decisions can help.

LOW-INPUT PRODUCTIVITY. Most heritage breeds of livestock are appreciated for the trait of productivity with limited inputs — a treasure that needs to be guarded. In brief, less is more: these animals will thrive with minimal inputs of concentrates, facilities, and health care.

LONGEVITY. In most heritage breeds, longevity of production is important. This means that offspring of older females should have some priority for retention into the herd. As a sort of coarse filter this one trait works well, because the older females have had many opportunities to be culled from the herd for disease, accident, unsoundness, or low production, and yet have proven to be worth keeping. Some caution is warranted, though, with waiting to select from only the very oldest, as their last few offspring frequently do not come up to the level of those produced a bit earlier in the dam's life.

Texas Longhorn cows are frequently fertile and productive up into their late teens, and sometimes beyond that. Capitalizing on that takes some selection for their offspring, so that the next generation has the same long productive life.

PEDIGREE. It can also be important to evaluate pedigrees, giving high priority to animals with rare bloodlines as long as other traits (such as production and breed type) are good. Pedigree selection by itself, however, can lead to mistakes if substandard or poorly performing animals are given high priority. Nevertheless, as a base for including candidates for selection into the breeding herd, a pedigree can be a useful tool to ensure that rare genetic influences are not lost to the herd and the breed.

GENERAL QUALITY. Finally, evaluating the conformation, growth, and production of the animals is critical. The ideal animal for retention has a balance of good conformation, has good growth or production of milk or fiber, has descended from a proven older dam, and has a bloodline that contributes genetic variation to the herd. Usually you must make some compromise among all of these, and in some cases extreme value in one dimension will argue for retention of an animal that may be weak in one of the other factors.

Texas Longhorn (CTLR) Cattle

An icon of the American Southwest, the Texas Longhorn is valued by ranchers for its ability to forage actively in arid environments, its high fertility, and its longevity. CTLR, the Cattlemen's Texas Longhorn Registry, is dedicated to preserving seed-stock Texas Longhorns.

Homeland: Texas and northeast Mexico

Traditional use: Beef production in arid landscapes

Size: 600–1,400 pounds

Colors: Highly variable colors

Other traits: Long horns

Special adaptations: Longevity, fertility in range conditions

Conservation status: Critical

Matchmaking Tools

Matching up animals to mate can be a tricky business, but fortunately breeders have a wide range of tools available to help them do this. Among these are pedigrees, production, and direct observation.

Pedigrees

Selection based on pedigrees is an especially important tool for very rare breeds. All breeders of heritage breeds need to pay attention to the genetic structure of their herds and flocks and also of the breed as a whole. The goal is to sample widely, and to make sure that sufficient genetic material gets passed along to the next generations so that inbreeding can be avoided.

Most genetic variation in most animal populations resides in the females. This basic principle is often overlooked. The old saying that "the herd sire is half the herd" can be only too accurate, and it refers to the overwhelming influence a heavily used sire can have throughout a herd. The saying reflects the fact that using a herd sire for several years results in a herd that is made up of his descendants, and no other influences. This has actually been a common occurrence in many isolated landrace herds, where a sire is used for long years and is then replaced by a son who repeats the pattern. Very soon the contributions of any other animals are minimized.

One way to avoid this problem is to ensure that herd sires are selected from different portions of the herd rather than repeatedly from only certain animals. A successful strategy, depending on herd size, is to use a sire for only one year, and then save a son for use in the next cycle. If this plan is followed with goats and sheep, and if the sires are used when yearlings, then the result is that every other year sees a different line of animals going through the herd. This broadens the sampling as much by years as by space (usually accomplished by having separate pastures) and can be an easy way to manage genetic diversity. In cattle, where bulls might be used as two-year-olds, it works even better because three sire lines are going through the herd just based on the timing of their use, the birth of their calves, and their dates of maturity and use of those calves. This manages the pedigrees before problems crop up!

More details of ways to secure the contributions of "rare pedigree" animals are outlined in chapter 8 in the sections on rescue projects.

RELATIVE GENETIC CONTRIBUTION TO THE HERD BY FOUNDER MALE AND FEMALES

Percentage of
Contribution of Sire

Percentage of
Contribution of Dams

USING A SINGLE SIRE FOR YEARS

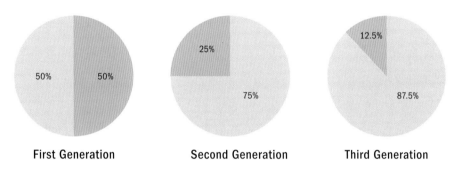

First Generation   Second Generation   Third Generation

USING SONS BACK ON FOUNDER FEMALES

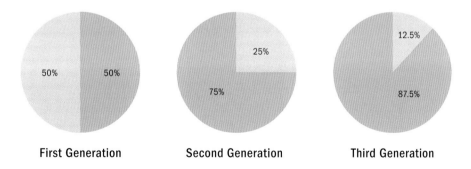

First Generation   Second Generation   Third Generation

Selection Based on Production

Animals vary in their levels of production, and a breeder's goal is to select the most productive individuals and recruit them into the breeding herd. *Production* has a range of meanings in different situations and for different breeds. The growth rate of a Devon beef animal on an improved pasture in the Midwest cannot be directly compared with that of a Texas Longhorn cow in arid West Texas. In each case, though, direct comparison of the animals for production traits can be done within the same breed, farm, and management system. Problems arise only when the comparison is across breeds, across farms, and across production systems.

Most important, "production" must be defined carefully. If production is only growth rate, then other traits such as disease-resistance and longevity may well suffer. Combining several traits into selection decisions makes for slower progress in any one trait but can work to conserve a useful balance of traits that will serve the breed, and its breeders, on into the future.

Selection Based on Observation

Nothing beats day-to-day contact with livestock in enabling you to evaluate temperament, ease of care, and other details that make animals either tough to be around or a delight. Watch the way they walk, observe how they interact with one another. Be sure to note which ones do their job in an efficient, easy way, and which ones are always causing problems of one sort or another.

Within the Sponenbergs' herd of Myotonic goats at Beechkeld Farm were three older sisters brought in from Texas. They formed a ruling triumvirate and were not to be messed with. Running the herd with an iron grip, they were nicknamed "the three hags" due to their management style (gratuitous violence). They were devoted to one another and hated all other goats!

Each of the three was large, smooth, and productive. Each had moments when she caused real management headaches. In this situation, close observation helps to make decisions, because an entire large herd of such characters would be impossible. Fortunately, in the case of these three, the high production passed on to the next generation, and they were all pacifists.

Poultry Selection Techniques

Poultry selection techniques mirror those of livestock. Longevity is less of an issue with most chickens, ducks, and turkeys but can still be important in some situations, particularly for geese. One reason for less concern over longevity is the higher reproductive rate of poultry, for the females lay many, many eggs in contrast to the few offspring per year of most livestock

REGAINING POULTRY SELECTION TECHNIQUES

Selection of poultry for productive characteristics has become a rare art in many areas as older masters pass away. In addition, people have become increasingly dependent on hatcheries for chicks rather than producing their own. The techniques of past master breeders have recently been rediscovered and are being used by an increasing number of modern producers. The Livestock Conservancy has helped to bridge this gap by training a new generation in traditional techniques during periodical hands-on workshops. The Livestock Conservancy additionally produces educational materials on breeding selection in chickens and turkeys.

Having more breeders trained in these selection strategies has dramatically improved the quality of birds in several breeds and has increased their productive potential in only a few generations. These materials serve as an important bridge from the past practices that assured the productivity of these breeds into a future where the breeds still have an important and useful role.

species. Egg production also tends to decline markedly with age, so that older animals are simply not productive enough to be economically viable.

Egg production and overall vigor are important traits in most poultry breeds. These can usually be evaluated by conformational traits such as depth of body, position of the bones around the vent, and other traits that are easier to measure than the individual egg production of females. The Livestock Conservancy has materials available on the details for these types of evaluations, and many breeders have successfully used these strategies to boost the productivity and beauty of their birds.

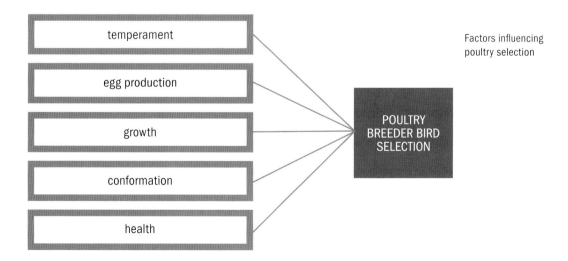

Factors influencing poultry selection

Relationships among Varieties in Chicken Breeds

Wyandotte chickens are wonderfully variable, having many different varieties (which are based on color) within the single breed (which is based on body size and style). Golden Laced and Silver Laced Wyandottes are among the more common of these, but the list also includes White, Black, Buff, Partridge, Silver Penciled, Columbian, and Blue Wyandottes. The Silver Laced, White, and Black form a trio of varieties with close genetic ties, and are different from one another in only the few genes that control these colors.

But the foundations of the other varieties come in part from a range of original sources. The Golden Laced variety has genes from the Cochin and Leghorn breeds; the Partridge Wyandotte has some Cochin and Cornish ancestors; and the Silver Penciled Wyandotte was developed by adding in some Brahma and Hamburg. The Columbian variety has added Plymouth Rock, and the Buffs were selected out of yet another mix that included Rhode Island Reds and Buff Cochins! In this breed, understanding the relationships between the varieties is important when mapping your breeding strategies. The varieties that are more closely related can easily be crossed among themselves. But including the more distantly related varieties into crosses only contributes to losing breed type and perfection of colors.

Superficially similar, Chantecler varieties have entirely different foundations.

In contrast, the two predominant varieties of Java (Black and Mottled) are close cousins and both have long histories within the Java breed. Other varieties, White and Auburn, also occasionally pop up from flocks of Black Javas as "sports." *Sport* here means an unusual color coming from the animals of the main breed. When sports are used for reproduction and are continued along they can frequently make new color varieties of an established breed.

Chantecler Chicken

The cold-hardy, dual-purpose Chantecler is an ideal choice for a northern small farm.

Homeland: Canada

Colors: Buff, white, partridge

Traditional uses: Eggs, meat

Other traits: Very small combs

Size: 6.5–9 pounds

Conservation status: Critical

Special adaptations: Cold-hardy

When color varieties come from sports, the varieties of a breed continue to be closely related, differing only in color. In contrast, varieties formed through crossbreeding and selection are less closely related.

As an example of this, the varieties of the Chantecler accepted by the American Poultry Association, White and Partridge, have little common genetic heritage, even though the general shape and function are similar across both. Each was developed from a different foundation: the White Chantecler from crosses of Dark Cornish, White Leghorn, Rhode Island Red, White Wyandotte, and White Plymouth Rocks; the Partridge from crosses of Partridge Wyandotte, Partridge Cochin, Dark Cornish, and Rose Comb Leghorn. These two varieties share little genetic background, despite being considered part of the same breed and having similar appearance and function.

The interrelationship of varieties should be carefully evaluated by breeders to ensure appropriate genetic management of poultry breeds. In some breeds, crosses between the varieties make good sense because they differ in only a few genes. In other breeds the varieties are much more like independent breeds with distinct histories and genetic backgrounds. Crossing varieties in this second sort of breed can set back a breeding program because it introduces too much genetic variation and makes the results unpredictable.

CHAPTER EIGHT

Rescues of Breeds, Strains, and Animals

FROM TIME TO TIME BREEDERS may encounter the last remnants of a breed or bloodline. In most cases, rescuing these few animals and expanding their numbers with careful attention to breeding strategies is possible. Examples of breeds being successfully saved from a very small remnant include Randall or Randall Lineback cattle, Ancona ducks, and several varieties of turkeys.

Strategies for Rescuing Breeds

SUCCESS WITH THESE IMPORTANT PROJECTS depends on having a good plan from the outset, and sticking to it.

Successful Breeds Move from Rescue to Production

Randall cattle, also known as Randall Lineback cattle, were brought back from the brink of extinction in the early 1980s (with fewer than 15 head). Today they number in the hundreds and benefit from having several breeders, each with a unique emphasis on breeding and management.

The breed was initially rescued by Cynthia Creech from the slaughterhouse door after she was alerted to the genetic uniqueness of the herd by The Livestock Conservancy. Cynthia, of Artemis Farm, used conservation breeding principles on this tiny remnant, including using multiple males and balancing the genetic contribution of founder animals. She was able to grow the breed from that original small and endangered remnant to a much larger population with a viable genetic structure.

Phil Lang, of Howland Homestead Farm, South Kent, Connecticut, helped by establishing one of the registries, and he also used the cows for milk production.

See page 28 for a Breed Snapshot of Randall Lineback cattle.

Joe Henderson's promotion of the breed is based on rose veal production (veal meat from older calves on a pasture-based system). Joe, of Chapel Hill Farm, Berryville, Virginia, has combined selection, commercial production, and attention to conservation in order to build strong demand for the breed and its products. This boost dramatically shifted the fate of the breed.

This diversity of approaches leads to increasing demand for the breed and has taken it from a population of fewer than 20 to its current level of several hundred.

Rescuing Small Populations

When a breed or bloodline has been whittled down to only a few animals, rescuing the population is best done by making sure all of the available females are mated each year. Ideally you would use different mates each time, because using a single male can create a genetic bottleneck, especially if he is used year after year after year.

Unfortunately, mating a rare group of females to a single male over several years is a common strategy. This practice is one of the main reasons why rare breeds and bloodlines decline in both vigor and productive potential. The formal term for this decline is *inbreeding depression*, which can occur slowly enough that by the time a breeder notices, it is too late to correct. Good breed stewardship demands that breeders pay close attention to the genetic structure of their herds to avoid situations where all animals are related. Once this happens, inbreeding cannot be avoided without outcrosses to other bloodlines.

Outright breed rescue is relatively rare, and let's hope that it remains that way! All the same, many breeds have a few animals with pedigrees and family relationships that are rare or unique. These animals should be carefully noted and documented as to pedigree and origin, and their role in the breeding program needs to be planned wisely. Strategically mating these animals can assure that their unique genetic contribution is not lost. Diluting valuable genetics with uncontrolled mating is all too common and can be avoided by careful planning.

So that inbreeding is not overwhelming, chart a rescue mating strategy carefully. Part of the key to success is to pay attention so that genetic material in the strain being rescued is not diluted to the degree that it has become meaningless for future use. If the genetic material is used well, it will continue to contribute to the breed.

RELATIVE GENETIC CONTRIBUTION TO THE HERD BY FOUNDER MALE AND FEMALES

Percentage of
Contribution of Sire

Percentage of
Contribution of Dams

USING ONE MALE FOR SEVERAL GENERATIONS

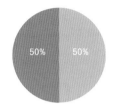

FIRST GENERATION
Bred from Original Sire & Original Dams

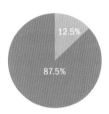

SECOND GENERATION
Same Sire Bred to Daughters

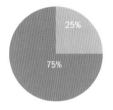

THIRD GENERATION
Same Sire Bred to Granddaughters

USING SONS OF ORIGINAL MALE FOR EACH GENERATION

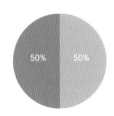

FIRST GENERATION
Sire Bred to Daughters of Original Dams

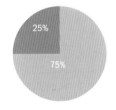

SECOND GENERATION
Son of Sire Bred to Original Dams

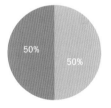

SECOND GENERATION
Son of Sire Bred to Daughters of Original Dams

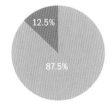

THIRD GENERATION
Grandson of Sire Bred to Original Dams

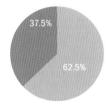

THIRD GENERATION
Grandson of Sire Bred to Daughters of Original Dams

An Exception to Every Rule

An exceptional Texas Longhorn cow is an example of the results of using a single bull over a herd for a long time. The cow's sire, grandsire, and great grandsire are all the same bull. The bull was used initially on a group of cows, then on his own daughters, and finally on his own granddaughters.

The cow was productive and fertile, which is far from usual after such close breeding. More common is for that strategy to result in inbreeding depression. This approach is not a wise breeding strategy for a heritage breed project.

In this case, as in most, the strategy was used for convenience rather than as a conscious choice. Extreme inbreeding like this does have a role in a few special situations, but it should always be done strategically and not from an absence of planning.

Rescue by Combining Diminished Strains

Many breeds with declining numbers end up in several different but isolated herds and flocks, and frequently each population is unrelated to the others. This occurrence is especially common in poultry breeds. In most cases the productivity of the animals has declined from inbreeding depression.

In this situation, when very few animals remain in a bloodline, saving the entire genetic package as an isolated unit is unlikely. The diminished bloodline must be combined with others to maintain their genetic contribution, ideally without losing the uniqueness that each brings to the breed.

One good strategy is to combine several rare strains of the breed, excluding from the mix any of the bloodlines that are more common, because the common bloodlines can dilute out the powerful contribution that can come from the rare bloodlines. The result of this type of combination is a *rare-bloodline composite*. These composites usually have both vigor and numerical strength from crossing the strains, and the advantage of being distantly related to the rest of the breed. The composite then provides a good choice for outcrossing to other more common strains. These animals become a useful and powerful resource for the future productivity and viability of the breed.

A note of caution: Though the initial recombination of the strains is likely to generate very vigorous animals that are close to the original heritage breed in conformation, type, and production, going forward from these initial recombined linecrossed animals can be tricky. Breeders must use careful thought and planning for the next generation of crosses rather than simply continuing with linecrosses to yet other strains. A more powerful genetic management program is to backcross to the first strains so that their contribution is concentrated.

Fred Diamond has saved remnants of several strains of Pineywoods cattle on his Mississippi farm by blending them into a composite herd.

See page 29 for a Breed Snapshot of Pineywoods cattle.

Rare-Strain Composites Provide Unique Opportunities for Breeders

Fred Diamond has successfully secured the survival of several old-strain remnants of the Pineywoods cattle breed and helped the breed's future by making these cattle available to other breeders. He has been diligent to use rare-strain bulls on cows of his herd's old family line, resulting in good genetic structure within the herd while maintaining a focus on rare genetics.

The herd is now a combination of Diamond, Seals, Bounds, and other family lines. All of these are otherwise extinct in the breed. Using rare strains together to make a composite has boosted the numbers, while producing cattle that are distinct from the more common (yet important!) strains of Conway, Baylis, Carter, and Hickman cattle in the breed.

Inbreeding Depression Is a Real Threat

Pineywoods cattle are a landrace breed, descended from Spanish stock, in the Pineywoods of the Gulf Coast states. One strain of Pineywoods cattle is the Palmer-Dunn strain. Muriel Dunn was the last owner, and this herd was a remnant of her family's once-numerous herd of cattle. At the time the bloodline was rescued, only a single old bull and eight cows were available. They were closely related to one another, as the practice in the herd was to use a single bull for several years running.

The cattle at the time of rescue were still vigorous and productive, but the next generation began to have the uneven reproductive performance that is typical of inbred animals. Some of the cows and bulls produced were sterile, others were still productive. To avoid a situation like this, heritage breed projects need to have careful planning and forethought. Unforeseen problems can easily arise and defeat success.

Fortunately, any problems with the Palmer-Dunn strain were caught in time. The original, fertile animals were still available, as well as some of their fully fertile offspring. Using these with each other as well as with other strains of the breed has allowed their strengths to persist in the breed, while bringing along none of their weaknesses. The Palmer-Dunn cattle were saved as a viable resource for Pineywoods cattle breeders seeking adapted, productive cattle.

Rescuing a Single Animal

The most drastic rescue is needed when only a single animal is available. When this individual is the last remnant of an entire breed, very little can be done. But, when a sole animal remains from a bloodline, a carefully coordinated breeding strategy can assure that this genetic material is not lost to the breed.

The "worst case" rescue is when only a single individual animal remains from an old bloodline. In this case, the best scenario is one in which this animal can produce offspring from a number of mates. This spreads the genetic material more widely in future generations. Careful attention to the contribution of the animal over several generations ensures that the genetic material is not lost. The easiest way to see if the genetic material is being lost is to examine pedigrees to determine if the animal's influence still persists. This step is especially important if the animal itself had superior performance or conformation. Ensuring that a male is mated for effective conservation is

This guinea Broadus/Griffen bull is stewarded by Billy Frank Brown and his son Jess, ensuring that this unique variant does not disappear.

easier than doing so for a female (a male animal can sire many offspring in a single breeding cycle, while a female can be a dam to only a single or single set of offspring), but various strategies can also be used to make certain that a lone female can make good contributions.

One effective strategy when only a single female is available is to mate her back to her own son. This type of mating produces offspring that are both her sons and her grandsons, which are therefore three-quarters her genetic influence, and only one-quarter the influence of the original son's sire. With luck, a male offspring results from this mating, and he can then sire many sons and daughters that will maintain the original female's genes in the population (they will be three-eighths her genetic influence). This strategy is tight inbreeding, but with care it can be useful in this situation, because the inbred animals can be outcrossed in the next generation, which effectively removes all of the inbreeding.

MATING A SINGLE FEMALE BACK TO HER OWN SONS

Percentage of Genetic
Contribution of Dam

Percentage of Genetic
Contribution of Males

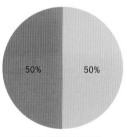

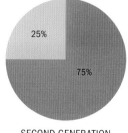

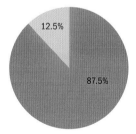

FIRST GENERATION
**Bred from Original Sire
and Original Dam**

SECOND GENERATION
**Original Dam Bred
to Son**

THIRD GENERATION
**Original Dam Bred to
Grandson**

Another strategy is to mate the female to different sires each year. Then her influence is represented in her several offspring, which are half-siblings. This broadens her genetic influence beyond the situation in which she is mated to the same sire year after year to produce full siblings. Because those full siblings bring along only a single male's influence in addition to hers, separating out her influence and genes from those of that single sire is difficult. Using multiple sires makes it possible to more adequately observe and concentrate her influence in future generations. This approach is also effective for daughters that result from mating a single female to her sons.

The two strategies (multiple sires, mating to sons) can be combined, though to what extent depends on the reproductive life of the female. With multiple sires producing multiple sons, breed the sons back to their mother so that three-quarter sons can be developed. In this case, the remaining one-quarter of the genetic material from these final sons comes from different original sires. This strategy provides some genetic distance even among these three-quarter siblings, and the end result is further distinction of the original dam's genetic material. Such matings also provide for a better and more complete sampling of the genetics of the original female. Each offspring of the original female has only one-half of her genetic influence, and those halves are different for each one. With more of those halves, more of the original genetic influence makes it to that next generation.

Mating a parent to an offspring is very close linebreeding, which is the mating of related individuals. Indeed, such a mating is so intense that it is inbreeding, which is the mating of very close relatives. This strategy does have very real risks; however, it has been used effectively in several breeds to salvage rare genetic material. Consider it a drastic sort of "first aid" for endangered genetics, and never use it as a long-term breeding strategy.

Rare Old Females Have a Unique Role

Flora Glendhu was a Leicester Longwool ewe that was brought to the farm at Colonial Williamsburg in Virginia. She was imported from Tasmania when Colonial Williamsburg reintroduced the breed to the United States, after a lapse of decades. She was the sole representative of her flock of origin. Fortunately, she also had good conformation, fertility, and a typical strong, lustrous longwool fleece.

The goal for Flora Glendhu was to maximize her contribution to the re-established breed in the United States. She was mated to an appropriate ram, and fortunately produced a son. That son was mated back to her, which is very close inbreeding. This mating produced a son/grandson, which was 75 percent her genetic influence. The son could be used more widely

Leicester Longwool Sheep

Both George Washington and Thomas Jefferson kept Leicester Longwools, a breed valued for its strong foraging ability, fertility, and heavy fleece of long, lustrous wool. The Colonial Williamsburg Foundation helped rescue the breed from near extinction in the 1980s. This breed is appropriate only for experienced sheep breeders.

Homeland: Leicestershire, England

Traditional uses: Wool, lamb production

Size: 125–250 pounds

Special adaptations: Foraging ability, heavy fleeces of lustrous long wool

Colors: White, gray, or black

Conservation status: Critical

to conserve her genetic influence in the breed than was possible with the single ewe.

This same strategy was followed twice, producing two of these "75 percent sons," both of which were then used widely in separate flocks. The breeding program also produced two "75 percent daughters," which were likewise used for breeding. Through these offspring, Glendhu's genetic material was used much more widely in the breed than would be possible for this single ewe from an old Tasmanian flock. Her genetics could easily have been lost to the breed without a specific strategy to assure their survival and use.

Breeder Commitment Is Key

Breeding animals is both an art and a science, and suffers if one or the other is missing. Careful attention to breeding produces breeds that are viable, useful, productive components of our agricultural system. Breeders have an essential role in this, as they each make individual selection decisions in their herds and flocks. This work is important, and many breeders make significant and long-lasting contributions to their breeds. A combination of dedication, observation, and skill is required, and then success is nearly guaranteed.

CHAPTER NINE

Boost Your Participation by Joining with Others

STRONG BREEDER COMMUNITIES are one of the main ways that heritage livestock and poultry breeds remain viable and relevant in agriculture. Associations, clubs, and other groups that organize around breeds can be wonderful examples of people joining together around a common goal of breed conservation. One of the most important functions of breed associations is to keep up with the census status and the genetic dynamics within the breed population so that bloodlines remain healthy and viable.

Joining a Breed Community

ASSOCIATIONS SERVE AS A CENTRAL PLACE for breeders to meet, learn, and discuss the breed. They educate all members, share techniques and new developments, and connect experienced breeders with the new and inexperienced ones. Associations are usually the first point of contact for new breeders or others interested in the animals, and can make a vital contribution to the health of the breed when it, and its members, refer prospective buyers of animals to others in their association. In the interests of the breed, breeders should be eager to do this when they do not have exactly what a buyer is looking for. This network encourages people to stay with the breed instead of choosing an alternative, and it makes a welcoming and supportive impression on both newcomers and fellow members.

ASSOCIATION MEETINGS HELP RARE BREED SURVIVAL

The annual meeting of the The Livestock Conservancy has become a vibrant forum for sharing information and networking in the interests of saving rare breeds. Participants bring their own stories of success and challenges with their heritage breed projects and, through exchanges with those involved in similar enterprises, take home new ideas to try out. Lectures and workshops offer opportunities for learning, which help attendees succeed with projects on their own farms and ranches. This annual meeting is but one example of the advantages of breeders getting together to exchange ideas and approaches and to encourage one another in the important task of saving heritage breeds.

Activities and Opportunities for All

Most breed associations organize a wide range of activities, including registration services for the breed, newsletters, shows and competitions, field days, award and recognition programs, and educational outreach programs.

These activities do not just happen, however; they require members who are willing to take on the various tasks. Figuring out how you can participate not only gives you an opportunity to do more for your breed but also puts you in touch with others who care and can be a resource for you. Some of the typical association tasks that you might be able to help with include the following.

WRITING FOR, OR EDITING AND PRODUCING THE NEWSLETTER. All newsletters need articles that are educational and appeal to the member community, and all editors welcome, at the very least, some occasional assistance. Newsletters are an essential part of breed association communication.

BEING THE REGISTRAR OR STUDBOOK KEEPER. This takes a great deal of time and commitment to update records in a timely and efficient manner. This key task should be undertaken only by those who can follow through on this large commitment.

ORGANIZING MEETINGS OR FIELD DAYS. These venues are important for developing the breed community into a strong and informed group.

VOLUNTEERING FOR ROUTINE TASKS AT MEETINGS, FIELD DAYS, OR SHOWS. This provides others with a break from sitting at information tables, or from other routine but vital functions that need to be done. A spin-off benefit is the chance to meet and interact with a wide variety of breeders.

One of the most productive things associations can do is to connect new with experienced members through mentoring, hands-on field days and trainings, and similar activities. In turn, younger breeders can assist older (and especially elderly) individuals and offer to help out with those tasks that get more difficult with age. This allows those important elderly breeders to stay active with the breed, and gives younger breeders the opportunity to learn successful strategies for breeding and managing animals.

Broad involvement of members in breeder associations does more to ensure long-term breed stability and survival than does any other single factor. Strong associations make sure that the breed is in good shape and is getting in the hands of new breeders. The best and strongest breed can have a weak breeder association, and this can lead to dwindling breed numbers and recruitment.

Newsletters Keep Members Engaged and Informed

Good, informative newsletters are an ideal way to distribute information about the breed and an especially easy way to communicate with folks inquiring at the breed office. Newsletters can be a healthy mix of educational articles, member profiles, and a forum for advertising that allows breeders to connect with potential customers.

For example, the Irish Draught Horse Society of North America publishes a quarterly newsletter, *The Blarney*. This engaging newsletter includes lots of photos of members and their horses, along with reports on recent and upcoming events and association business. It also includes articles from members on everything from health concerns to training issues to discussions of breed standards and selection criteria. *The Blarney* is a treasure trove of information about the status of the breed as a whole, providing society members with up-to-date herdbook, census, distribution information, and valuable insights and news from the Irish homeland of the breed.

The Blarney newsletter

Field Days Help Everyone!

Leicester Longwool sheep breeders in the United States have a long tradition of annual meetings, which offer educational seminars as well as hands-on evaluation of sheep. Additional field days that are regional or even more local allow the more experienced individuals to share their knowledge with the less experienced ones, and raise the general ability of everyone involved to evaluate sheep as candidates for breeding stock.

These types of events help to cement personal relationships and contacts among breeders. The meetings are especially helpful in providing a platform for varying views of the breed to be expressed and discussed. Breeders communicate their hopes and aspirations, and that helps the Leicester Longwool move into the future.

Shared Experience and Wisdom Help Breeds

Millie and Dave Holderread carefully steward many breeds of ducks and geese and, just as importantly, new generations of waterfowl breeders.

Dave Holderread exemplifies the impact that a single dedicated and accomplished breeder can have on an association as well as on animals. His work in streamlining the American Poultry Association standards for several breeds of waterfowl has ensured a secure future for many heritage breeds. His eager sharing of his techniques and knowledge through his excellent handbooks for breeders has helped many new enthusiasts get started. Dave is also known as a source for excellent breeding fowl among serious breeders. Whole-hearted participation by talented folks like Dave assures that breeds remain strong.

Breeds Need New as Well as Traditional Breeders

For more than two centuries, Spanish goats were the primary meat goat in the United States. Their fate took a turn for the worse when Boer goats were imported from South Africa in 1993. Many herds of Spanish goats were crossbred with Boer bucks, effectively eliminating the purebreds in a single generation. Fortunately for Spanish goats, several traditional breeders are still in Texas with herds of up to 1,000 head or more.

A number of dedicated and enthusiastic younger breeders are eager to pursue a range of projects with the breed. Some of these new breeders are focusing on commercial production, and others on the conservation and rescue of the rarest of the bloodlines of this once-common breed. Young and old breeders are working together for the future of the breed because they realize that it takes a variety of approaches and attitudes to effectively conserve a landrace breed.

When Breed Associations Go Bad

UNFORTUNATELY, DIVISIONS AMONG MEMBERS can be responsible for the complete breakdown of an association and a good breeding program. Often these divisions spring from differences of opinion between two important groups: traditionalists with long experience in the breed, and newcomers who come with fresh ideas and approaches.

Many heritage breeds have a long legacy of traditional husbandry techniques and attitudes, which older breeders grew up with and have guarded jealously over their long lives. Old-timers need to be respected for their long contributions to breed definition and survival.

Without newcomers, though, heritage breeds are doomed. Newcomers frequently come to a breed community with great enthusiasm. They also

Spanish Goat

Southwestern strains of the Spanish goat are adapted to arid environments; in the Southeast they are adapted to humid places with high parasite loads.

Homeland: Texas; other strains from Southeast and California

Traditional use: Meat production

Size: 60–175 pounds

Colors: Vary widely

Other traits: Most Spanish goats are horned

Conservation status: Watch

have at least some knowledge of approaches and directions that would help the animal population expand numbers and market share. These can change traditional ways, and often strike traditionalists as dangerous to the breed, its history, and especially to the role of tradition in the community.

Breeder communities need both. The well-experienced individuals bring continuity (the breed's heritage) while the novices bring future potential (legacy moving forward). Or to put it another way, without the old-timers the breed would never have survived in the first place, while without the new recruits it will have nowhere to go next. Each group has to make sure that the community is working successfully for the other group, and both groups will typically have to make occasional compromises so that the entire community supporting the breed can move forward.

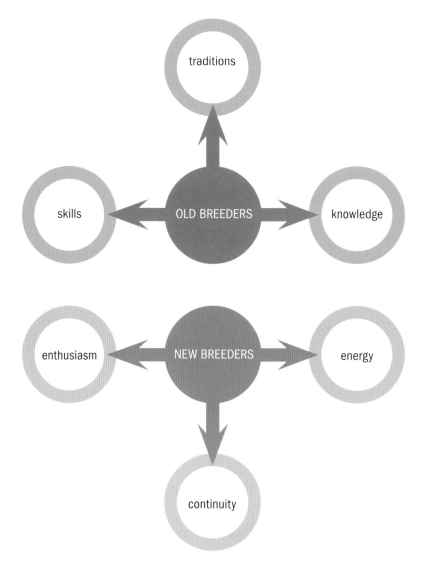

Old and new breeders bring different strengths to breeds.

Local Breeds Have Local Characters

Landraces, or local breeds, face special challenges when a fiercely loyal group of traditional older breeders have kept the breed going but now finds the need to become more formally organized in order to secure the breed's future. Often these owners have strong opinions, and their track record proves that those opinions have merit. The crisis arises as this group retires or passes on but unfortunately lacks a next generation to keep the livestock heritage of the family intact.

This circumstance makes it necessary to actively engage a younger generation of outsiders. These newcomers often have different attitudes and may emphasize different aspects of the breed. Fortunately, some breed associations have been able to merge people with disparate opinions into a successful and vibrant group to keep the breed's fate secure. Active programs for the Milking Devon breed, for example, have been able to successfully blend traditional approaches with those of the next generation, and the diversity of strategies for breeding and managing the cattle is reflected in the composition of the Association's board of directors.

Some landrace breeds have special challenges in this regard. Some have pedigreed herds, but many do not because the animals were (and are) raised on large ranges, in huge herds with multiple sires. While individual pedigrees are impossible in this situation, at least the purebred status of the animals is certain when only a single breed runs the range, as was typical for most of the history of these important breeds.

Challenges for this group of breeds encompass the entire range of conservation challenges including breed definition, animal identification, organized breed associations, and registry management. Traditional practices do not always address these challenges very well, but new practices tend to be rejected by older traditional breeders.

Knowing Your Exit Strategy

All conservation breeders, if they truly care about the future of their chosen breed, should have a good exit strategy. No one starts a heritage breed project with the idea of getting out any time soon, but the fact is that the average length of involvement for breeders is about five years. When you factor in those breeders who have put in 60 years or more, that "five-year average" covers a great many projects that are short indeed! Far too often a breed undergoes huge and important losses when producers buy up uniquely bred animals and then hastily disperse them after a few years of involvement.

Great damage to a breed is also done when, as has happened all too often in the annals of heritage breed history, an experienced breeder dies or quits and the animals are abruptly taken to the local sale yard. Whatever good work that producer achieved over many years is lost on that very day. This needs to be avoided at all costs.

Three good ways to avoid such losses are:

- having breeders within a breed community who consistently and eagerly share genetic material with other breeders through regular sales

- assuring that final dispersals are to other stewards and not only to the sale yards. This can be by mentioning animals in a will, and arranging for them to go to an informed and careful steward or to the breed association.

- maintaining your records and your relationships with other breeders so that when the time to retire from a project arrives, you and your stock are a known quantity that others will willingly promote and purchase.

All rare breed stewards should plan early and well for the animals in their care to be put in the hands of other conservation-minded individuals should the need arise.

Fitting Your Project into the Larger Breed Picture

As a steward of a heritage breed, especially if that breed is endangered, you have a responsibility to ensure that your animals contribute to the health of the overall population by understanding how your animals' genetics fit into those of the overall breed. This aspect of rare breed management can be extremely important when others holding the same bloodline lose animals, or lose entire herds due to unforeseen trouble. This subtlety can make all the difference to a rare breed and its future.

Outlining the pitfalls of a heritage breed endeavor always runs the risk of making it sound like a daunting, challenging task. The reality is that the pitfalls are few, and all can be managed by just a little bit of planning and forethought. The rewards are many. Your participation in a vibrant group doing important conservation work can make that work enjoyable for all. Your participation also can make lasting positive contributions that are rewarding day-to-day as you work with the animals, and also in the long term as you reflect on the real and permanent progress those contributions have made to your chosen breed.

Heritage Breeds: A Future for You and for Agriculture

Heritage breeds "work" only if they are a bridge between the past and the present. They need to survive the present in order to be available to serve the future. This endeavor is exciting and requires that new and enthusiastic breeders are recruited along the way.

Heritage breeds, beyond being a significant part of our agricultural past as well as our agricultural future, are fun and rewarding. The rewards are many, including the animals' beauty and compelling history, their potential to provide for farm income, and their role in providing options to farmers both now and in the future.

To best survive, these breeds need owners who are sensitive to tradition as well as to the present and future potential of these breeds to be vital and productive parts of America's agricultural landscape. A huge part of this process is economic opportunity, because breeds that secure their owners an economic return can be assured of not only survival but also expansion and security.

Contributions to heritage breed survival are, and should always remain, enjoyable, culturally significant, and also a key component of agricultural security for our nation. All of this rests on the contribution of individuals, each working diligently to save and steward a small part of this important whole.

APPENDIX

Breeds at a Glance

E ACH SPECIES HAS A WONDERFULLY DIVERSE ARRAY of breeds. The summaries below barely scratch the surface of all the details that make up each of them. Use these tables as a general guideline, and then go out and meet the animals and their breeders in person to discover which breed appeals to you most and fits best with your goals.

Rabbits

The ideal temperature for rabbits is about 50°F (10°C).

threatened

American

Origin: United States
Purpose: Meat, fur
Adult Weight M/F (lbs.): 9–11/10–12
Color: Blue, white
Litter Size: 8–10
Temperament: Docile

Mothering Ability: Good
Owner Skill: Novice
Notes & Tips: Select for good mandolin body shape. The blue variety is the deepest blue color of all rabbits.

critical

American Chinchilla

Origin: United States
Purpose: Meat, fur
Adult Weight M/F (lbs.): 9–11/10–12
Color: Chinchilla
Litter Size: 8–10

Temperament: Docile
Mothering Ability: Good
Owner Skill: Novice
Notes & Tips: Fast growth; good meat-to-bone ratio

threatened

Belgian Hare

Origin: Belgium
Purpose: Exhibition
Adult Weight M/F (lbs.): 6–9.5/6–9.5
Color: Brown with black ticking
Litter Size: 4–8
Temperament: Active, intelligent

Mothering Ability: Variable
Owner Skill: Intermediate to advanced
Notes & Tips: Created to resemble a
wild hare but is a true rabbit. Can be a
challenge to breed.

watch

Beveren

Origin: Belgium
Purpose: Meat, fur
Adult Weight M/F (lbs.): 8–11/9–12
Color: Black, blue, white
Litter Size: 8–12
Temperament: Docile

Mothering Ability: Good
Owner Skill: Novice
Notes & Tips: Select for mandolin
body shape. Good cold tolerance. Quick
growth rate.

threatened

Blanc de Hotot

Origin: France
Purpose: Meat, fur
Adult Weight M/F (lbs.): 8–11/9–11
Color: Frosty white with thin black eye
bands
Litter Size: 6–8

Temperament: Docile to active
Mothering Ability: Good to Variable
Owner Skill: Novice to intermediate
Notes & Tips: Meat lighter in color than
other breeds

recovering

Crème d'Argent

Origin: France
Purpose: Meat, fur
Adult Weight M/F (lbs.): 8–10.5/8.5–11
Color: Creamy white with orange cast,
bright orange undercoat
Litter Size: 5–8

Temperament: Docile
Mothering Ability: Good
Owner Skill: Novice
Notes & Tips: Known to have relaxed
personality. Ears should be well rounded
at end.

watch

Giant Chinchilla

Origin: United States
Purpose: Meat, fur
Adult Weight M/F (lbs.): 12–15/13–16
Color: Chinchilla
Litter Size: 7–8
Temperament: Docile

Mothering Ability: Good
Owner Skill: Novice
Notes & Tips: Requires platform in cage
to avoid sore hocks. Quick growth rate;
large size requires bigger cages.

study

Harlequin

Origin: France

Purpose: Exhibition, meat

Adult Weight M/F (lbs.): 6.5–9/7–9.5

Color: Black, blue, chocolate, lilac

Litter Size: 3–5

Temperament: Docile

Mothering Ability: Good, are known to foster kits as well

Owner Skill: Novice to intermediate

Notes & Tips: Mostly kept for exhibition but size produces a decent carcass for meat. Color patterns hard to perfect.

watch

Lilac

Origin: British

Purpose: Meat, fur

Adult Weight M/F (lbs.): 5.5–7.5/6–8

Color: Dove gray with pink tint

Litter Size: 4–6

Temperament: Docile

Mothering Ability: Variable

Owner Skill: Novice

Notes & Tips: Does better in cooler climates. Slow to mature but hardy.

watch

Rhinelander

Origin: Germany

Purpose: Exhibition

Adult Weight M/F (lbs.): 6.5–9.5/7–10

Color: Black and orange or blue and fawn

Litter Size: 10

Temperament: Active but docile, intelligent

Mothering Ability: Good to variable

Owner Skill: Intermediate

Notes & Tips: Color pattern a challenge to perfect. Owners train rabbits to run and pose for the show table.

threatened

Silver

Origin: Portugal, possibly early Thai breed before then

Purpose: Exhibition

Adult Weight M/F (lbs.): 4–7/4–7

Color: Black, brown, or fawn with silver ticking

Litter Size: 3–6

Temperament: Docile but active

Mothering Ability: Variable

Owner Skill: Novice

Notes & Tips: Small size makes them easy to handle

threatened

Silver Fox

Origin: United States

Purpose: Meat, fur

Adult Weight M/F (lbs.): 9–11/10–12

Color: Black or blue with silver ticking

Litter Size: 7–8

Temperament: Docile

Mothering Ability: Good

Owner Skill: Novice

Notes & Tips: Fast growth. Fur stands upright and does not fall back like other rabbit coats.

Chickens

watch

Ancona

Origin: Italy
Purpose: Eggs
Adult Weight M/F (lbs.): 6/4.5
Egg Color: White
Egg Size: Medium to large
Rate of Lay: Good (120–180)
Temperament: Highly active, pheasant-like

Brooding & Mothering: Non-setters
Owner Skill: Novice
Ideal Climate: Does well in both cold and heat. Single comb vulnerable to frostbite.
Notes & Tips: Noted for hardiness and vigor. Prefers free range over wide areas. Good winter layers.

study

Araucana

Origin: South American multiple-country composite
Purpose: Eggs
Adult Weight M/F (lbs.): 5/4
Egg Color: Blue
Egg Size: Medium to large
Rate of Lay: Good (about 150)
Temperament: Docile

Brooding & Mothering: Setters
Owner Skill: Novice to intermediate
Ideal Climate: Does well in both cold and heat
Notes & Tips: Genetic fertility issues can be a challenge for breeding. Hardy breed.

watch

Aseel

Origin: India and Pakistan
Purpose: Meat
Adult Weight M/F (lbs.): 5.5/4
Egg Color: White or tinted light brown
Egg Size: Small
Rate of Lay: Poor (6–40)
Temperament: Both sexes can be highly aggressive even at a very young age
Brooding & Mothering: Setters; exceedingly protective mothers

Owner Skill: Intermediate to advanced
Ideal Climate: Best in warm climates; can tolerate some cold but needs to be kept dry
Notes & Tips: Makes excellent crosses for the production of broilers. Very hardy and predator-savvy.

Australorp

recovering

Origin: Australia
Purpose: Eggs, meat
Adult Weight M/F (lbs.): 8.5/6.5
Egg Color: Brown
Egg Size: Large
Rate of Lay: Excellent (250–250)
Temperament: Active, yet gentle

Brooding & Mothering: Setters
Owner Skill: Novice
Ideal Climate: Does well in both cold and heat
Notes & Tips: Productive and fast growing. A favorite for pastured egg production.

Blue Andalusian

threatened

Origin: Spain
Purpose: Eggs
Adult Weight M/F (lbs.): 7/5.5
Egg Color: Chalk-white
Egg Size: Large
Rate of Lay: Very good (about 160)
Temperament: Active, yet gentle
Brooding & Mothering: Non-setters

Owner Skill: Novice to intermediate
Ideal Climate: Best in hot to moderate climates
Notes & Tips: Very rugged and robust. Blue feather coloration can be a challenge to perfect. Excels in free-range conditions. Typically not good in confinement.

Brahma

watch

Origin: United States
Purpose: Meat, eggs
Adult Weight M/F (lbs.): 12/9.5
Egg Color: Brown
Egg Size: Medium to large
Rate of Lay: Good (140)
Temperament: Docile
Brooding & Mothering: Setters

Owner Skill: Novice to intermediate
Ideal Climate: Does better in cold climates
Notes & Tips: Good winter layers. Best kept on well-drained soils. Perches should be no more than 12 inches off the ground due to breed's body size.

Buckeye

threatened

Origin: United States
Purpose: Meat, eggs
Adult Weight M/F (lbs.): 9/6.5
Egg Color: Brown
Egg Size: Large
Rate of Lay: Good (120–150)
Temperament: Active, yet gentle, very curious
Brooding & Mothering: Variable

Owner Skill: Novice
Ideal Climate: Does well in the cold but can adapt to heat over time
Notes & Tips: Excellent broilers. Breast meat is almost as dark as thighs. Best ranger of the American class.

threatened

Buttercup

Origin: Sicily

Purpose: Eggs

Adult Weight M/F (lbs.): 6.5/5

Egg Color: White

Egg Size: Small to medium

Rate of Lay: Good (140–180)

Temperament: Highly active

Brooding & Mothering: Non-setters

Owner Skill: Intermediate

Ideal Climate: Best in hot to moderate climates

Notes & Tips: Good forager and does well on free range. Striking flower-shaped comb that can be a challenge to perfect.

critical

Campine

Origin: Belgium

Purpose: Eggs

Adult Weight M/F (lbs.): 6/4

Egg Color: White

Egg Size: Medium to large

Rate of Lay: Good (150)

Temperament: Very active

Brooding & Mothering: Non-setters

Owner Skill: Novice

Ideal Climate: Best in hot to moderate climates

Notes & Tips: Vigorous forager. Can be sexed as day-olds when silver variety hens are crossed with golden variety roosters.

watch

Catalana

Origin: Spain

Purpose: Eggs, meat

Adult Weight M/F (lbs.): 8/6

Egg Color: White to tinted

Egg Size: Medium

Rate of Lay: Good (150)

Temperament: Active, can be flighty but not overly so

Brooding & Mothering: Non-setters

Owner Skill: Novice

Ideal Climate: Best in hot to moderate climates

Notes & Tips: Succulent meat. Cockerels often used for capon in Spain.

critical

Chantecler

Origin: Canada

Purpose: Eggs, meat

Adult Weight M/F (lbs.): 8.5/6.5

Egg Color: Brown

Egg Size: Large

Rate of Lay: Good (120–180)

Temperament: Docile but some lines can be variable

Brooding & Mothering: Variable

Owner Skill: Novice

Ideal Climate: Best in cold climates; not recommended for hot regions

Notes & Tips: Good winter layer. Almost no wattles and a tiny button comb. Well-fleshed breast.

watch

Cochin

Origin: China
Purpose: Meat, ornamental
Adult Weight M/F (lbs.): 11/8.5
Egg Color: Brown
Egg Size: Large
Rate of Lay: Fair (up to 140)
Temperament: Docile

Brooding & Mothering: Setters
Owner Skill: Novice
Ideal Climate: Best in cold climates
Notes & Tips: Excellent broody breed — especially for turkey and duck eggs. Needs to be kept on well-drained soils.

watch

Cornish

Origin: England
Purpose: Meat
Adult Weight M/F (lbs.): 10.5/8
Egg Color: Tinted
Egg Size: Medium to large
Rate of Lay: Fair (50–80)
Temperament: Docile
Brooding & Mothering: Non-setters

Owner Skill: Intermediate
Ideal Climate: Best in warm climates; does not do well in the cold or extreme heat
Notes & Tips: Superb table bird. Very short compact body with enormous breast.

critical

Crevecour

Origin: France
Purpose: Meat, eggs
Adult Weight M/F (lbs.): 8/6.5
Egg Color: White
Egg Size: Medium to large
Rate of Lay: Good (about 150)
Temperament: Active
Brooding & Mothering: Non-setters
Owner Skill: Intermediate

Ideal Climate: Does not do well in weather extremes; best in temperate climate
Notes & Tips: Known for fine flesh. Has fine bones and excellent percentage of meat to offal when processed. Can be delicate to manage. Vulnerable to predators due to crest of feathers on head obstructing vision.

threatened

Cubalaya

Origin: Cuba
Purpose: Eggs, meat
Adult Weight M/F (lbs.): 6/4
Egg Color: White
Egg Size: Small
Rate of Lay: Good (125–175)
Temperament: Active but calm
Brooding & Mothering: Setters

Owner Skill: Novice to intermediate
Ideal Climate: Best in hot to moderate climates; highly tolerant of humidity
Notes & Tips: Known for fine meat qualities. Can be aggressive to other birds but mild-mannered compared to other game birds. Can be noisy.

critical

Delaware

Origin: United States
Purpose: Eggs, meat
Adult Weight M/F (lbs.): 8.5/6.5
Egg Color: Brown
Egg Size: Large to jumbo
Rate of Lay: Very good (150–200)
Temperament: Docile

Brooding & Mothering: Setters
Owner Skill: Novice
Ideal Climate: Does well in both hot and cold climates
Notes & Tips: Developed to grow quickly for broiler production. Color pattern can be a challenge to perfect.

watch

Dominique

Origin: United States
Purpose: Eggs, meat
Adult Weight M/F (lbs.): 7/5
Egg Color: Brown
Egg Size: Medium
Rate of Lay: Excellent (230–275)
Temperament: Active

Brooding & Mothering: Variable
Owner Skill: Novice
Ideal Climate: Does well in any climate
Notes & Tips: Excellent forager and homestead fowl. Primarily a laying fowl but has nice breast proportions for meat.

threatened

Dorking

Origin: England
Purpose: Eggs, meat
Adult Weight M/F (lbs.): 9/7
Egg Color: White
Egg Size: Medium to large
Rate of Lay: Good (about 150)
Temperament: Docile but active

Brooding & Mothering: Setters
Owner Skill: Novice
Ideal Climate: Does well in any climate, but comb can be prone to frostbite
Notes & Tips: Famous for meat flavor. Good winter layers and foragers.

threatened

Faverolles

Origin: France
Purpose: Meat, eggs
Adult Weight M/F (lbs.): 8/6.5
Egg Color: Tinted to light brown
Egg Size: Medium to large
Rate of Lay: Very good (180–200)
Temperament: Docile but active
Brooding & Mothering: Variable
Owner Skill: Novice

Ideal Climate: Can do well in any climate, but comb can be prone to frostbite
Notes & Tips: Used for meat and winter egg production. Productive when in contact with the ground. Does not like confinement.

watch

Hamburg

Origin: Holland
Purpose: Eggs
Adult Weight M/F (lbs.): 5/4
Egg Color: White
Egg Size: Medium
Rate of Lay: Excellent (200–225)
Temperament: Active, can be very flighty

Brooding & Mothering: Non-setters
Owner Skill: Novice to intermediate
Ideal Climate: Does well in any climate
Notes & Tips: Likes to roost in trees. Lays large numbers of eggs over a longer period of lifetime than most breeds.

critical

Holland

Origin: United States
Purpose: Eggs, meat
Adult Weight M/F (lbs.): 8.5/6.5
Egg Color: White
Egg Size: Large
Rate of Lay: Good
Temperament: Docile

Brooding & Mothering: Setters
Owner Skill: Novice
Ideal Climate: Does well in both hot and cold, but comb can be prone to frostbite
Notes & Tips: Top-notch foragers; good homestead fowl

watch

Houdans

Origin: France
Purpose: Meat, eggs
Adult Weight M/F (lbs.): 8/6.5
Egg Color: White
Egg Size: Small to medium
Rate of Lay: Fair
Temperament: Docile, exceptionally gentle
Brooding & Mothering: Setters, but can be awkward and break eggs

Owner Skill: Novice to intermediate
Ideal Climate: Does well in both hot and cold climates but prefers drier climates
Notes & Tips: A five-toed fowl. "Label Rouge" certified in France for flavor. Can be vulnerable to predators due to crest of feathers on head obstructing vision.

threatened

Java

Origin: United States
Purpose: Meat, some eggs
Adult Weight M/F (lbs.): 9.5/7.5
Egg Color: Brown
Egg Size: Large
Rate of Lay: Good (150)
Temperament: Docile but active

Brooding & Mothering: Setters
Owner Skill: Novice
Ideal Climate: Does well in both hot and cold but combs may be prone to frostbite
Notes & Tips: Premier homestead fowl. Excellent forager. Slower growth equating to excellent meat flavor.

watch

Jersey Giant

Origin: United States

Purpose: Meat, eggs

Adult Weight M/F (lbs.): 13/10

Egg Color: Brown

Egg Size: Extra large

Rate of Lay: Very good (175–185)

Temperament: Docile

Brooding & Mothering: Setters, but can be awkward and break eggs

Owner Skill: Novice to intermediate

Ideal Climate: Will do better in cooler climates

Notes & Tips: Envisioned to take the place of the turkey on the table

watch

La Fleche

Origin: France

Purpose: Meat, eggs

Adult Weight M/F (lbs.): 8/6.5

Egg Color: White

Egg Size: Medium to large

Rate of Lay: Very good (200)

Temperament: Active

Brooding & Mothering: Non-setters

Owner Skill: Intermediate

Ideal Climate: Can handle hot and cold but not extremes of either

Notes & Tips: Highly regarded for flavor of the meat. Large breast for size of bird.

threatened

Lakenvelder

Origin: Holland & Germany

Purpose: Eggs

Adult Weight M/F (lbs.): 5/4

Egg Color: White

Egg Size: Medium

Rate of Lay: Very good (150–200)

Temperament: Active, can be flighty

Brooding & Mothering: Non-setters

Owner Skill: Novice

Ideal Climate: Can handle hot and cold but combs can be prone to frostbite

Notes & Tips: Excellent foragers; very predator savvy

threatened

Langshan

Origin: China

Purpose: Eggs, meat

Adult Weight M/F (lbs.): 9.5/7.5

Egg Color: Very dark brown with purple tint

Egg Size: Large

Rate of Lay: Good (historically up to 200)

Temperament: Docile

Brooding & Mothering: Setters

Owner Skill: Novice

Ideal Climate: Good in both hot and cold; only Asiatic breed suited for warm climates

Notes & Tips: Produces large amount of very white meat

recovering

Leghorn

Origin: Italy
Purpose: Eggs
Adult Weight M/F (lbs.): 6/4.5
Egg Color: White
Egg Size: Medium to large
Rate of Lay: Excellent (250–300)
Temperament: Very active

Brooding & Mothering: Non-setters
Owner Skill: Novice to intermediate
Ideal Climate: Best in hot to moderate climates
Notes & Tips: Noted for hardiness and vigor; produces more eggs with less food than any other breed

threatened

Malay

Origin: India
Purpose: Meat
Adult Weight M/F (lbs.): 9/7
Egg Color: Brown
Egg Size: Medium
Rate of Lay: Poor
Temperament: Active. Can be quarrelsome in confinement, will feather-pick each other in close quarters.

Brooding & Mothering: Setters. Cannot cover many eggs due to tight feathering. Males can be aggressive to chicks.
Owner Skill: Intermediate
Ideal Climate: Best in hot to moderate climates
Notes & Tips: Tallest of all chickens. Meat is very lean with little fat. Adults are hardy but chicks can be delicate.

study

Manx Rumpy (Persian Rumpless)

Origin: Iraq
Purpose: Eggs, meat
Adult Weight M/F (lbs.): 6/5.5
Egg Color: Tinted to brown
Egg Size: Medium
Rate of Lay: Good
Temperament: Active; roosters can be slightly aggressive to new arrivals in flock

Brooding & Mothering: Setters
Owner Skill: Intermediate
Ideal Climate: Good in both hot and cold, but combs may be prone to frostbite
Notes & Tips: Exceptional free-range bird with little supplement needed. Fertility tends to be low.

watch

Minorca

Origin: Spain
Purpose: Eggs
Adult Weight M/F (lbs.): 9/7.5
Egg Color: Chalk-white
Egg Size: Extra large
Rate of Lay: Excellent
Temperament: Very active
Brooding & Mothering: Non-setters

Owner Skill: Novice
Ideal Climate: Best in hot to moderate climates
Notes & Tips: Great forager. Produces large carcass but meat can be dry. Historically, breasts were stuffed with lard before being roasted.

critical

Modern Game

Origin: England

Purpose: Show, meat

Adult Weight M/F (lbs.): 6/4.5

Egg Color: White to tinted light brown

Egg Size: Small

Rate of Lay: Fair (100)

Temperament: Active and curious

Brooding & Mothering: Setters

Owner Skill: Intermediate

Ideal Climate: Best in hot to moderate climates

Notes & Tips: Doesn't tolerate close confinement. Can be aggressive at times. Noisy. Needs lots of exercise to maintain muscle tone.

critical

Nankin

Origin: England

Purpose: Eggs, broody hens

Adult Weight M/F (lbs.): 1.5/1.4 (bantam breed)

Egg Color: Creamy white

Egg Size: Small

Rate of Lay: Fair (90–100)

Temperament: Docile

Brooding & Mothering: Setters; hens are excellent broodies

Owner Skill: Novice

Ideal Climate: Best in hot to moderate climates; rose comb variety better for cold

Notes & Tips: Eggs tend to be round in shape, making them a challenge to hatch. They typically hatch better under a broody hen, not an incubator.

watch

New Hampshire

Origin: United States

Purpose: Eggs, meat

Adult Weight M/F (lbs.): 8.5/6.5

Egg Color: Brown

Egg Size: Extra large

Rate of Lay: Good (160–180)

Temperament: Docile

Brooding & Mothering: Variable

Owner Skill: Novice

Ideal Climate: Does well in hot and cold but combs can be prone to frostbite

Notes & Tips: Rapid rate of growth; fine table fowl

watch

Old English Game

Origin: England

Purpose: Show

Adult Weight M/F (lbs.): 5/4

Egg Color: White to tinted light brown

Egg Size: Small

Rate of Lay: Fair (120)

Temperament: Active, can be aggressive

Brooding & Mothering: Setters; can be distracted mothers

Owner Skill: Intermediate to advance

Ideal Climate: Best in hot to moderate climates

Notes & Tips: Can be capable of long flight and can revert to feral state

Orpington

Origin: England

Purpose: Meat, eggs

Adult Weight M/F (lbs.): 10/8

Egg Color: Brown

Egg Size: Large to extra large

Rate of Lay: Very good (175–200)

Temperament: Docile with friendly disposition

Brooding & Mothering: Setters

Owner Skill: Novice

Ideal Climate: Best in cooler climates due to large body size but can acclimate to warmer regions

Notes & Tips: Excellent rate of growth in some lines. Fine table bird.

Phoenix

Origin: Germany

Purpose: exhibition, feathers

Adult Weight M/F (lbs.): 5.5/4

Egg Color: White to tinted brown

Egg Size: Small

Rate of Lay: Poor

Temperament: Docile

Brooding & Mothering: Setters

Owner Skill: Novice to intermediate

Ideal Climate: Best in hot to moderate climates

Notes & Tips: Highly ornamental with extraordinarily long tail. Chicks need extra protein in diet when growing their tails. Need lots of room to roam and clean facilities to keep tails in good condition.

Plymouth Rock

Origin: United States

Purpose: Eggs, meat

Adult Weight M/F (lbs.): 9.5/7.5

Egg Color: Brown

Egg Size: Large

Rate of Lay: Very good (about 200)

Temperament: Docile

Brooding & Mothering: Setters

Owner Skill: Novice

Ideal Climate: Does well in both hot and cold, but combs can be prone to frostbite in extreme cold

Notes & Tips: Early feathering and rapid growth rate. Good forager.

Polish

Origin: Holland

Purpose: Eggs

Adult Weight M/F (lbs.): 6/4.5

Egg Color: White

Egg Size: Medium to large

Rate of Lay: Very good (200)

Temperament: Docile

Brooding & Mothering: Non-setters

Owner Skill: Intermediate

Ideal Climate: Can handle hot and cold but not extremes

Notes & Tips: Dense crest on head can obscure vision. Flighty when disturbed. Raising chicks in elevated brooders so they can see human activity makes them calmer. Vulnerable to predators.

critical

Redcap

Origin: England
Purpose: Eggs
Adult Weight M/F (lbs.): 7.5/6
Egg Color: White
Egg Size: Medium to large
Rate of Lay: Very good (180–220)
Temperament: Active

Brooding & Mothering: Non-setters
Owner Skill: Novice
Ideal Climate: Best in hot to moderate climates
Notes & Tips: Huge rose comb. Adult color not fully developed until the second or third year.

recovering

Rhode Island Red

Origin: United States
Purpose: Eggs, meat
Adult Weight M/F (lbs.): 8.5/6.5
Egg Color: Brown
Egg Size: Large
Rate of Lay: Excellent (250)
Temperament: Docile

Brooding & Mothering: Variable
Owner Skill: Novice
Ideal Climate: Does well in both hot and cold climates; rose comb variety better choice for cold than single comb
Notes & Tips: Once considered finest-flavored American chicken

watch

Rhode Island White

Origin: United States
Purpose: Eggs, meat
Adult Weight M/F (lbs.): 8.5/6.5
Egg Color: Brown
Egg Size: Large to extra large
Rate of Lay: Good
Temperament: Docile; roosters can be housed together in winter but will need more room at the start of breeding season

Brooding & Mothering: Non-setters
Owner Skill: Novice
Ideal Climate: Does well in both hot and cold climates
Notes & Tips: Good forager and loves to range but not far from the coop

critical

Russian Orloff

Origin: Russia
Purpose: Meat
Adult Weight M/F (lbs.): 8.5/6.5
Egg Color: Light brown
Egg Size: Small
Rate of Lay: Variable

Temperament: Can be aggressive
Brooding & Mothering: Non-setters
Owner Skill: Intermediate to advanced
Ideal Climate: Excellent cold tolerance; not recommended for hot climates
Notes & Tips: Very hardy breed

watch

Sebright

Origin: England
Purpose: Eggs, show
Adult Weight M/F (lbs.): 1.4/1.25
(bantam breed)
Egg Color: White or creamy
Egg Size: Very small
Rate of Lay: Poor
Temperament: Docile

Brooding & Mothering: Non-setters
Owner Skill: Novice to intermediate
Ideal Climate: Can handle hot and cold
but not extremes; can be delicate
Notes & Tips: Primarily an exhibition
fowl. Great choice if you don't have a lot
of room for chickens.

watch

Shamo

Origin: Japan
Purpose: Meat, ornamental
Adult Weight M/F (lbs.): 11/7
Egg Color: Pale brown
Egg Size: Medium
Rate of Lay: Fair (60–100)
Temperament: Can be aggressive to
other birds but friendly to people

Brooding & Mothering: Setters
Owner Skill: Intermediate to advanced
Ideal Climate: Best in hot to moderate
climates
Notes & Tips: Meat firm and almost
tough. Meat often used in Sumo
wrestler's diet.

critical

Spanish

Origin: Spain
Purpose: Eggs
Adult Weight M/F (lbs.): 8/6.5
Egg Color: Chalk-white
Egg Size: Large
Rate of Lay: Good
Temperament: Active; chicks can be
flighty but adults tend to be more calm
and curious

Brooding & Mothering: Non-setters
Owner Skill: Novice
Ideal Climate: Best in hot to moderate
climates
Notes & Tips: Known as the clown-
faced chicken due to the big white face
patches of the roosters.

critical

Sultan

Origin: Turkey
Purpose: Ornamental
Adult Weight M/F (lbs.): 6/4
Egg Color: White
Egg Size: Small
Rate of Lay: Good (150–200)
Temperament: Active

Brooding & Mothering: Variable
Owner Skill: Novice to intermediate
Ideal Climate: Best in hot to moderate
climates
Notes & Tips: Can be less damaging to
vegetation than other breeds. Chicks can
be delicate. Has large breast and fine meat.

critical

Sumatra

Origin: Sumatra, Java, Borneo
Purpose: Ornamental
Adult Weight M/F (lbs.): 5/4
Egg Color: White to tinted
Egg Size: Small
Rate of Lay: Fair
Temperament: Active

Brooding & Mothering: Setters
Owner Skill: Novice to intermediate
Ideal Climate: Best in hot to moderate climates
Notes & Tips: Lustrous plumage. Flighty and jumpy. Good at evading predators.

recovering

Sussex

Origin: England
Purpose: Meat, eggs
Adult Weight M/F (lbs.): 9/7
Egg Color: Tan to brown
Egg Size: Large
Rate of Lay: Very good (about 200), if hens are not overly fat

Temperament: Docile, curious
Brooding & Mothering: Setters
Owner Skill: Novice
Ideal Climate: Good in hot and cold, but combs can be prone to frostbite
Notes & Tips: Good all-around table bird. Fattens easily. Good ranger.

recovering

Wyandotte

Origin: United States
Purpose: Eggs, meat
Adult Weight M/F (lbs.): 8.5/6.5
Egg Color: Brown
Egg Size: Large
Rate of Lay: Very good (180–200)
Temperament: Docile

Brooding & Mothering: Variable
Owner Skill: Novice
Ideal Climate: Excellent for cold but can do well in hot
Notes & Tips: Great dual-purpose bird for homesteads

critical

Yokohama

Origin: Germany
Purpose: Ornamental
Adult Weight M/F (lbs.): 4.5/3.5
Egg Color: Tinted light brown
Egg Size: Small
Rate of Lay: Poor to fair (80); known to go broody after laying only 12–14 eggs
Temperament: Active
Brooding & Mothering: Variable

Owner Skill: Novice to intermediate
Ideal Climate: Best in hot to moderate climates
Notes & Tips: An exhibition fowl. Chicks need extra protein when growing their long tail feathers. Needs lots of room to roam.

Ducks

critical

Ancona

Origin: Unknown, likely North America
Purpose: Meat, eggs
Adult Weight M/F (lbs.): 6.5/6
Egg Color: White, tinted, blue-green, or spotted
Egg Size: Large
Rate of Lay: Excellent (210–280)

Temperament: Docile, active
Brooding & Mothering: Fair to good
Owner Skill: Novice
Ideal Climate: Adaptable
Foraging Ability: Excellent
Notes & Tips: Rapid growth. Excellent for slug control.

study

Australian Spotted

Origin: United States
Purpose: Exhibition
Adult Weight M/F (lbs.): 2.2/2
Egg Color: Cream, blue, green
Egg Size: Small to medium
Rate of Lay: Poor to fair (50–125)
Temperament: Docile

Brooding & Mothering: Excellent
Owner Skill: Novice
Ideal Climate: Adaptable
Foraging Ability: Excellent
Notes & Tips: Can fly but typically stays near home. Great for insect control.

critical

Aylesbury — traditional

Origin: England
Purpose: Meat
Adult Weight M/F (lbs.): 10/9
Egg Color: White or tinted green
Egg Size: Extra large
Rate of Lay: Poor to fair (35–125)
Temperament: Docile
Brooding & Mothering: Poor to fair
Owner Skill: Intermediate

Ideal Climate: Adaptable
Foraging Ability: Fair
Notes & Tips: Good meat-to-bone ratio. Pink bills will turn orange over time with UV exposure. Traditional type can be challenging to breed. Need pool to mate properly.

Buff (Orpington)

Origin: England
Purpose: Meat, eggs
Adult Weight M/F (lbs.): 8/7
Egg Color: White or tinted
Egg Size: Large
Rate of Lay: Very good (150–220)
Temperament: Docile, active

Brooding & Mothering: Fair to good
Owner Skill: Novice to intermediate
Ideal Climate: Adaptable
Foraging Ability: Good
Notes & Tips: Light pin feathers make for a cleaner dress-out

Campbell

Origin: England
Purpose: Eggs
Adult Weight M/F (lbs.): 4.5/4
Egg Color: White to tinted
Egg Size: Large
Rate of Lay: Excellent (250–340)

Temperament: Docile, active
Brooding & Mothering: Poor to fair
Owner Skill: Novice to intermediate
Ideal Climate: Adaptable
Foraging Ability: Excellent
Notes & Tips: Premier egg-layers

Cayuga

Origin: United States
Purpose: Meat, eggs
Adult Weight M/F (lbs.): 8/7
Egg Color: Black fading to blue-gray and then white by end of lay season
Egg Size: Large
Rate of Lay: Good (100–150)
Temperament: Docile

Brooding & Mothering: Good
Owner Skill: Novice
Ideal Climate: Adaptable
Foraging Ability: Good
Notes & Tips: Many skin this duck rather than pluck because of dark pin feathers

Dutch Hookbill

Origin: Holland
Purpose: Eggs
Adult Weight M/F (lbs.): 4/3.5
Egg Color: White to blue-green
Egg Size: Large
Rate of Lay: Variable (100–225)
Temperament: Docile, active

Brooding & Mothering: Good
Owner Skill: Novice to intermediate
Ideal Climate: Adaptable
Foraging Ability: Excellent
Notes & Tips: One of the best foraging ducks. Can fly.

Magpie

Origin: Wales

Purpose: Meat, eggs

Adult Weight M/F (lbs.): 6/5.5

Egg Color: White

Egg Size: Medium to large

Rate of Lay: Excellent (220–290)

Temperament: Docile, active, can be high-strung

Brooding & Mothering: Fair to good

Owner Skill: Intermediate

Ideal Climate: Adaptable

Foraging Ability: Excellent

Notes & Tips: One of the best foraging ducks. Strong sex drive, so you should have no fewer than 5 ducks per drake.

Rouen — traditional

Origin: France

Purpose: Meat

Adult Weight M/F (lbs.): 10/8

Egg Color: White

Egg Size: Extra large

Rate of Lay: Poor to fair (30–125)

Temperament: Docile

Brooding & Mothering: Poor to good

Owner Skill: Intermediate

Ideal Climate: Adaptable

Foraging Ability: Good

Notes & Tips: Excellent table bird. Needs pool to mate properly. Can be a challenge to breed.

Runner

Origin: Malasia

Purpose: Eggs

Adult Weight M/F (lbs.): 4.5/4

Egg Color: White to blue-green

Egg Size: Large

Rate of Lay: Very good (200)

Temperament: Docile

Brooding & Mothering: Poor to fair

Owner Skill: Novice to intermediate

Ideal Climate: Adaptable

Foraging Ability: Excellent

Notes & Tips: Often used for herding-dog training and trials

Saxony

Origin: Germany

Purpose: Meat, eggs

Adult Weight M/F (lbs.): 9/8

Egg Color: White to blue-green

Egg Size: Extra large

Rate of Lay: Very good (190–240)

Temperament: Docile

Brooding & Mothering: Fair to good

Owner Skill: Novice

Ideal Climate: Adaptable

Foraging Ability: Good

Notes & Tips: One of the best dual-purpose breeds. Slower growing with lean meat.

Silver Appleyard

critical

Origin: England

Purpose: Meat, eggs

Adult Weight M/F (lbs.): 9/8

Egg Color: White

Egg Size: Large to extra large

Rate of Lay: Excellent (200–270)

Temperament: Docile

Brooding & Mothering: Fair to good

Owner Skill: Novice

Ideal Climate: Adaptable

Foraging Ability: Good

Notes & Tips: Has deep wide breast

Swedish

watch

Origin: Pomerania (currently Germany & Poland)

Purpose: Meat, eggs

Adult Weight M/F (lbs.): 8/6.5

Egg Color: White, green, tinted

Egg Size: Large

Rate of Lay: Good (100–150)

Temperament: Docile, active

Brooding & Mothering: Fair to good

Owner Skill: Intermediate

Ideal Climate: Adaptable

Foraging Ability: Good to excellent

Notes & Tips: Does not do well in confinement. Color can be a challenge to perfect.

Welsh Harlequin

critical

Origin: Wales

Purpose: Meat, eggs

Adult Weight M/F (lbs.): 5.5/5

Egg Color: White to tinted

Egg Size: Large

Rate of Lay: Excellent (240–330)

Temperament: Docile, active

Brooding & Mothering: Poor to good

Owner Skill: Novice

Ideal Climate: Adaptable

Foraging Ability: Excellent

Notes & Tips: Carcass dresses very clean. Great dual-purpose bird.

Geese

Geese are adaptable to both hot and cold climates.

watch

African

Origin: Asia
Purpose: Meat
Adult Weight M/F (lbs.): 20/18
Color: Gray/brown, buff, white
Egg Color: White
Egg Size: Extra large
Rate of Lay: 20–45

Temperament: Some males aggressive
Brooding & Mothering: Fair to good
Owner Skill: Novice to intermediate
Notes & Tips: Takes several years to fully mature. Needs shelter in cold to prevent frostbite on knob.

critical

American Buff

Origin: United States
Purpose: Meat
Adult Weight M/F (lbs.): 18/16
Color: Buff
Egg Color: White
Egg Size: Large

Rate of Lay: 25–35
Temperament: Docile, curious
Brooding & Mothering: Good
Owner Skill: Novice
Notes & Tips: Not as noisy as some other goose breeds

watch

Chinese

Origin: Asia
Purpose: Meat, eggs
Adult Weight M/F (lbs.): 12/10
Color: Brown, white
Egg Color: White
Egg Size: Medium to large
Rate of Lay: 40–100
Temperament: Most are docile; some can be aggressive

Brooding & Mothering: Poor to fair
Owner Skill: Novice
Notes & Tips: Make good "watchdogs." Good for small properties. Need shelter in cold to prevent frostbite on knob.

critical

Cotton Patch

Origin: United States
Purpose: Meat, weeding
Adult Weight M/F (lbs.): 14/12
Color: White (male), gray or saddleback (female)
Egg Color: White
Egg Size: Large

Rate of Lay: 12–16 eggs/year (6–8 eggs/clutch, twice a year)
Temperament: Docile, active
Brooding & Mothering: Good
Owner Skill: Intermediate
Notes & Tips: Most strains retain the ability to fly

critical

Pilgrim

Origin: United States
Purpose: Meat
Adult Weight M/F (lbs.): 14/13
Color: White (male), gray (female)
Egg Color: White
Egg Size: Large

Rate of Lay: 25–40
Temperament: Docile
Brooding & Mothering: Good
Owner Skill: Novice
Notes & Tips: Fairly peaceful birds. Goslings can be sexed as day-olds.

critical

Pomeranian

Origin: Germany
Purpose: Meat
Adult Weight M/F (lbs.): 17/15
Color: Gray or buff, saddleback, white
Egg Color: White
Egg Size: Large

Rate of Lay: 25–40
Temperament: Some males aggressive
Brooding & Mothering: Good
Owner Skill: Novice to intermediate
Notes & Tips: Color can be a challenge to perfect. Make good "watchdogs."

critical

Roman

Origin: Italy
Purpose: Meat, guarding
Adult Weight M/F (lbs.): 12/10
Color: White
Egg Color: White
Egg Size: Large
Rate of Lay: 25–35

Temperament: Some males aggressive
Brooding & Mothering: Good
Owner Skill: Novice to intermediate
Notes & Tips: Make good "watchdogs"

threatened

Sebastapol

Origin: Southeastern Europe

Purpose: Ornamental

Adult Weight M/F (lbs.): 14/12

Color: White, gray, buff

Egg Color: White

Egg Size: Large

Rate of Lay: 25–35

Temperament: Docile

Brooding & Mothering: Good to excellent

Owner Skill: Novice to intermediate

Notes & Tips: Up to 5 geese per gander

study

Steinbacher

Origin: Germany

Purpose: Fighting, meat

Adult Weight M/F (lbs.): 14/12

Color: Gray, blue, buff

Egg Color: White

Egg Size: Large

Rate of Lay: 12–15

Temperament: Can have goose-to-goose aggression; females very defensive of young

Brooding & Mothering: Variable to good

Owner Skill: Intermediate

Notes & Tips: Only the blue color variety found in the United States

watch

Toulouse (Dewlap)

Origin: France

Purpose: Meat

Adult Weight M/F (lbs.): 26/20

Color: Gray, buff

Egg Color: White

Egg Size: Extra large

Rate of Lay: 20–35, few 60

Temperament: Docile

Brooding & Mothering: Poor to fair

Owner Skill: Intermediate to advanced

Notes & Tips: Relatively quiet breed. Matures slowly. Challenging to breed. Needs at least 15"-deep pool to mate in. In heat will need water and shade.

Turkeys

Turkeys can live in most climates but skin on head and neck can be prone to frostbite in extreme cold. For large birds, heat stress can occur, so ensuring they have access to shade and ample water will be key to managing birds in hot climates.

critical

Beltsville Small White

Origin: United States
Purpose: Meat
Adult Weight M/F (lbs.): 21/12

Owner Skill: Novice
Notes & Tips: Difficult to source; dresses clean

watch

Black

Origin: Europe
Purpose: Meat
Adult Weight M/F (lbs.): 33/18
Owner Skill: Novice

Notes & Tips: May have dark pin feathers or ink spots on skin when processed

watch

Bourbon Red

Origin: United States
Purpose: Meat
Adult Weight M/F (lbs.): 33/18
Owner Skill: Novice to intermediate

Notes & Tips: Dresses clean. Deep mahogany red color can be difficult to perfect. Avoid reddish orange birds.

watch

Standard Bronze

Origin: United States
Purpose: Meat
Adult Weight M/F (lbs.): 36/20
Owner Skill: Novice to intermediate

Notes & Tips: May have dark pin feathers. A very challenging color pattern to perfect.

critical

Chocolate

Origin: United States
Purpose: Meat
Adult Weight M/F (lbs.): 33/18

Owner Skill: Novice
Notes & Tips: Dresses clean

critical

Jersey Buff

Origin: United States
Purpose: Meat
Adult Weight M/F (lbs.): 33/18

Owner Skill: Novice
Notes & Tips: Dresses clean

critical

Lavender (Lilac)

Origin: United States
Purpose: Meat
Adult Weight M/F (lbs.): 33/18

Owner Skill: Novice to intermediate
Notes & Tips: Dresses clean. Color can be difficult to perfect.

critical

Midget White

Origin: United States
Purpose: Meat, eggs
Adult Weight M/F (lbs.): 20/10
Owner Skill: Novice

Notes & Tips: Dresses clean. Exceptional layer among turkeys (up to 80 eggs). Many are very people-oriented birds.

threatened

Narragansett

Origin: United States
Purpose: Meat
Adult Weight M/F (lbs.): 33/18

Owner Skill: Novice to intermediate
Notes & Tips: May have dark pin feathers. Difficult color to perfect.

watch

Royal Palm

Origin: United States
Purpose: Meat, exhibition
Adult Weight M/F (lbs.): 22/12
Owner Skill: Novice to intermediate

Notes & Tips: Seems to be somewhat more heat tolerant than other strains. Difficult feather pattern to perfect.

watch

Slate

Origin: United States
Purpose: Meat
Adult Weight M/F (lbs.): 33/18

Owner Skill: Novice to intermediate
Notes & Tips: May have dark pin feathers. Difficult color to perfect.

threatened

White Holland

Origin: United States
Purpose: Meat
Adult Weight M/F (lbs.): 36/20

Owner Skill: Novice
Notes & Tips: Dresses clean

Sheep

Primary purpose is listed; however, wool from meat breeds is sold to crafters, and meat from wool breeds can be very flavorful.

recovering

Barbados Blackbelly

Origin: Caribbean
Purpose: Meat
Weight (M/F, lbs.): 100–130/85–100
Average Number of Lambs: 2.5
Temperament: Alert but docile

Owner Skill: Novice to intermediate
Foraging Ability: Excellent
Ideal Climate: Hot and humid
Notes & Tips: Prolific. Polled. Hair sheep. Parasite resistant.

threatened

Black Welsh Mountain

Origin: Wales
Purpose: Wool
Weight (M/F, lbs.): 130/100–110
Average Number of Lambs: 1.75
Temperament: Variable by flock

Owner Skill: Novice
Foraging Ability: Good
Ideal Climate: Cold and damp
Notes & Tips: The wool is true black and not a deep brown. Lambs are vigorous.

threatened

Clun Forest

Origin: England
Purpose: Meat, wool
Weight (M/F, lbs.): 200/150
Average Number of Lambs: 1.75
Temperament: Docile

Owner Skill: Novice
Foraging Ability: Excellent
Ideal Climate: Temperate
Notes & Tips: Hardy, self-sufficient breed with versatile wool

threatened

Cotswold

Origin: England
Purpose: Meat, wool
Weight (M/F, lbs.): 250/175
Average Number of Lambs: 2
Temperament: Docile
Owner Skill: Novice to intermediate
Foraging Ability: Good

Ideal Climate: Temperate
Notes & Tips: Longwool breed. Strong mothering ability.

threatened

Dorset Horn

Origin: England
Purpose: Meat, wool
Weight (M/F, lbs.): 200/160
Average Number of Lambs: 2
Temperament: Docile

Owner Skill: Novice to intermediate
Foraging Ability: Average
Ideal Climate: Temperate
Notes & Tips: Breed out of season

critical

Florida Cracker

Origin: United States
Purpose: Meat
Weight (M/F, lbs.): 125–175/100–125
Average Number of Lambs: 1.25
Temperament: Alert but docile

Owner Skill: Novice to intermediate
Foraging Ability: Excellent
Ideal Climate: Hot and humid
Notes & Tips: Parasite resistant

critical

Gulf Coast

Origin: United States
Purpose: Meat
Weight (M/F, lbs.): 150–190/100–140
Average Number of Lambs: 1.25
Temperament: Alert but docile
Owner Skill: Novice to intermediate

Foraging Ability: Excellent
Ideal Climate: Hot and humid
Notes & Tips: Horned or polled; parasite resistant; white, gray, or black wool

critical

Hog Island

Origin: United States
Purpose: Meat
Weight (M/F, lbs.): 125–150/90–100
Average Number of Lambs: 1.5
Temperament: Alert but docile

Owner Skill: Novice to intermediate
Foraging Ability: Excellent
Ideal Climate: Temperate to hot and humid
Notes & Tips: Horned or polled

threatened

Jacob — American

Origin: England, then United States
Purpose: Wool
Weight (M/F, lbs.): 160/110
Average Number of Lambs: 1.5
Temperament: Docile
Owner Skill: Novice to intermediate

Foraging Ability: Good
Ideal Climate: Temperate
Notes & Tips: Two, four, or occasionally six horns; spotted black and white

threatened

Karakul — American

Origin: Iran, then United States
Purpose: Meat, wool, some dairy
Weight (M/F, lbs.): 175–225/100–110
Average Number of Lambs: 1
Temperament: Docile

Owner Skill: Novice to intermediate
Foraging Ability: Good
Ideal Climate: Cold and dry
Notes & Tips: Fat-rumped; horned.
Black, brown, white, or gray.

critical

Leicester Longwool

Origin: England
Purpose: Wool, meat
Weight (M/F, lbs.): 250/180
Growth Rate: Average
Average Number of Lambs: 2
Temperament: Docile

Owner Skill: Advanced
Foraging Ability: Good
Ideal Climate: Temperate
Notes & Tips: Large. Slow-growing.
Luster longwool. Mothering ability good
once bonded with lambs.

watch

Lincoln

Origin: England
Purpose: Wool, meat
Weight (M/F, lbs.): 250–300/200–275
Average Number of Lambs: 2
Temperament: Docile

Owner Skill: Intermediate
Foraging Ability: Good
Ideal Climate: Temperate
Notes & Tips: Large. Slow-growing.
Luster longwool.

threatened

Navajo-Churro

Origin: United States
Purpose: Wool, meat
Weight (M/F, lbs.): 120–200/85–120
Average Number of Lambs: 1.25
Temperament: Alert but docile

Owner Skill: Novice to intermediate
Foraging Ability: Excellent
Ideal Climate: Cold to hot, and dry
Notes & Tips: Horned or polled, some
with four horns. Desert adapted. Prolific.

watch

Oxford

Origin: England
Purpose: Meat, wool
Weight (M/F, lbs.): 250–350/200–250
Average Number of Lambs: 2
Temperament: Docile
Owner Skill: Novice to intermediate

Foraging Ability: Good
Ideal Climate: Temperate
Notes & Tips: Large meek sheep; sire
breed

critical

Romeldale/CVM

Origin: United States
Purpose: Wool, meat
Weight (M/F, lbs.): 275/160
Average Number of Lambs: 1.75
Temperament: Docile

Owner Skill: Novice
Foraging Ability: Good
Ideal Climate: Hot and dry
Notes & Tips: Prolific; long lived; wide range of colors

critical

Santa Cruz

Origin: United States
Purpose: Wool
Weight (M/F, lbs.): 100–150/80–100
Average Number of Lambs: 1.25
Temperament: Alert

Owner Skill: Intermediate
Foraging Ability: Excellent
Ideal Climate: Hot and dry
Notes & Tips: Originally a free-ranging sheep with good survival traits

threatened

St. Croix

Origin: Caribbean
Purpose: Meat
Weight (M/F, lbs.): 165–200/125–150
Average Number of Lambs: 1.75
Temperament: Docile

Owner Skill: Novice
Foraging Ability: Excellent
Ideal Climate: Hot and humid
Notes & Tips: Prolific. White polled hair sheep breed

recovering

Shetland

Origin: Shetland Islands
Purpose: Wool, meat
Weight (M/F, lbs.): 110/85
Average Number of Lambs: 1.5
Temperament: Docile

Owner Skill: Novice
Foraging Ability: Excellent
Ideal Climate: Cold and damp
Notes & Tips: Versatile multicolored wool

watch

Shropshire

Origin: England
Purpose: Meat, some dairy, wool
Weight (M/F, lbs.): 240/170
Growth Rate: Rapid
Average Number of Lambs: 2
Temperament: Docile

Owner Skill: Novice
Foraging Ability: Good
Ideal Climate: Temperate
Notes & Tips: Long-lived; often has twins or triplets

recovering

Southdown

Origin: England
Purpose: Meat, wool
Weight (M/F, lbs.): 200/150
Average Number of Lambs: 2
Temperament: Docile

Owner Skill: Novice
Foraging Ability: Good
Ideal Climate: Temperate
Notes & Tips: Very thrifty; affectionate

watch

Tunis

Origin: United States
Purpose: Meat, wool
Weight (M/F, lbs.): 200–275/125–175
Average Number of Lambs: 2
Temperament: Docile

Owner Skill: Novice
Foraging Ability: Good
Ideal Climate: Temperate
Notes & Tips: Polled. Southern-adapted. Prolific.

recovering

Wiltshire Horn

Origin: England
Purpose: Meat
Weight (M/F, lbs.): 230/175
Average Number of Lambs: 2
Temperament: Variable

Owner Skill: Novice to intermediate
Foraging Ability: Good
Ideal Climate: Hot and dry
Notes & Tips: Horned white hair sheep

Goats

critical

Arapawa

Origin: New Zealand
Purpose: Meat, some dairy
Adult Weight M/F (lbs.): 125/60–80
Color: Tan, black, or striped
Temperament: Active breed, can tame down nicely with handling early in life

Owner Skill: Novice to intermediate
Ideal Climate: Does well in cooler climates
Notes & Tips: Rugged; excellent mothering ability

study

Golden Guernsey

Origin: England
Purpose: Dairy
Adult Weight M/F (lbs.):
190–200/125–150
Color: Gold, shaggy

Owner Skill: Novice to intermediate
Temperament: Docile, friendly
Notes & Tips: Produces up to 2,000 lbs milk per year. Milk has high fat content.

recovering

Myotonic (Tennessee Fainting)

Origin: United States
Purpose: Meat
Adult Weight M/F (lbs.):
150–175/50–150
Color: Variable; usually black and white
Temperament: Docile

Owner Skill: Novice to intermediate
Ideal Climate: Does well in both hot and cold
Notes & Tips: Heavy muscling. Excellent mothering ability. Parasite resistance.

recovering

Oberhasli

Origin: Switzerland
Purpose: Dairy
Adult Weight M/F (lbs.):
125–150/100–125
Color: Tan with black trim, rarely black or red

Temperament: Docile; does can be pushy with each other at times
Owner Skill: Novice to intermediate
Notes & Tips: Up to 4,665 lbs milk per year (record). Are said to be less fearful of water and trail obstacles than other goats.

critical

San Clemente

Origin: United States

Purpose: Meat

Adult Weight M/F (lbs.):
80–120/50–80

Color: Tan and black with face stripes

Temperament: Active breed; can tame down nicely with handling early in life

Owner Skill: Novice to intermediate

Ideal Climate: Can adapt to both warm and cold

Notes & Tips: Source locally so they are adapted to local parasites. Excellent mothering ability.

watch

Spanish

Origin: United States

Purpose: Meat and some dairy

Adult Weight M/F (lbs.): 200/50–125

Color: Variable

Temperament: Active and rugged

Owner Skill: Novice to intermediate

Notes & Tips: Long-lived. Fertile. Rugged. Excellent mothering ability.

Pigs

critical

Choctaw

Origin: Oklahoma
Purpose: Lard, meat
Mature Weight (M/F, lbs.):
250–300/150–200
Color: Black; rarely, other colors
Litter Size: 6 or more

Temperament: Alert to aggressive
Owner Skill: Intermediate
Notes & Tips: Free-ranging hogs from the Choctaw Nation. Adaptable and independent.

critical

Gloucestershire Old Spots

Origin: England
Purpose: Meat, lard, bacon
Mature Weight (M/F, lbs.):
500–600/450–500
Color: White with black spots
Litter Size: 10 or more

Temperament: Docile
Owner Skill: Beginner to intermediate
Notes & Tips: Lop-eared orchard pigs. Hardy. Sows need 14 teats for those large litters.

critical

Guinea Hog

Origin: Southern United States
Purpose: Lard, meat
Mature Weight (M/F, lbs.):
250–300/150
Color: Black; rarely, red or blue

Litter Size: 4–8
Temperament: Docile
Owner Skill: Beginner
Notes & Tips: Erect ears. Calm. Self-sufficient. Good mothers.

watch

Hereford

Origin: United States
Purpose: Meat, bacon
Mature Weight (M/F, lbs.): 800/600
Color: Red and white

Litter Size: 6 or more
Temperament: Docile
Owner Skill: Beginner
Notes & Tips: Well muscled, lean type

critical

Large Black

Origin: England
Purpose: Bacon, lard, meat
Mature Weight (M/F, lbs.):
700–800/500–600
Color: Black
Litter Size: 8–10 or more

Temperament: Docile
Owner Skill: Beginner to intermediate
Notes & Tips: Lop-eared; hardy

critical

Mulefoot

Origin: United States
Purpose: Lard, meat
Mature Weight (M/F, lbs.): 600/400
Color: Black
Litter Size: 5–8

Temperament: Active but docile
Owner Skill: Beginner
Notes & Tips: Erect to semi-lop ears;
forage well

critical

Ossabaw Island

Origin: United States
Purpose: Lard, bacon
Mature Weight (M/F, lbs.):
250–350/150–250
Color: Black; black and tan spotted

Litter Size: 4–8
Temperament: Active but docile
Owner Skill: Beginner to intermediate
Notes & Tips: Thrifty and self-sufficient

critical

Red Wattle

Origin: United States
Purpose: Meat, bacon
Mature Weight (M/F, lbs.):
700–800/500–600
Color: Red

Litter Size: 10–15
Temperament: Docile
Owner Skill: Beginner
Notes & Tips: Good foraging ability;
fertile; good mothers

threatened

Tamworth

Origin: Ireland
Purpose: Bacon, meat
Mature Weight (M/F, lbs.):
500–600/500–600
Color: Red

Litter Size: 6–10
Temperament: Active but docile
Owner Skill: Beginner
Notes & Tips: Erect ears; long lean
body; good foragers

Cattle

threatened

Ancient White Park

Origin: England
Purpose: Beef
Size (M/F, lbs.): 1,000–1,300/800–1,000
Color: White with black points
Owner Skill: Advanced

Temperament: Can be aggressive
Foraging Ability: Excellent
Ideal Climate: Temperate
Notes & Tips: Good choice in regions with large predators

recovering

Ankole-Watusi

Origin: Central Africa
Purpose: Dual
Size (M/F, lbs.): 1,000–1,600/900–1,200
Color: Usually red or red and white
Owner Skill: Novice to intermediate

Temperament: Docile
Foraging Ability: Excellent
Ideal Climate: Hot and dry
Notes & Tips: African cattle used for beef production. Efficient; huge horns and gentle nature.

watch

Ayrshire

Origin: Scotland
Purpose: Dairy
Size (M/F, lbs.): 1,600/1,100
Color: Red and white
Owner Skill: Novice to intermediate

Temperament: Assertive
Foraging Ability: Average
Ideal Climate: Temperate
Notes & Tips: Elegant dairy cattle with long lives. Forage well.

recovering

Belted Galloway

Origin: Scotland
Purpose: Beef
Size (M/F, lbs.): 1,600/1,000
Color: White belt on black, red, or dun
Owner Skill: Novice to intermediate

Temperament: Docile to assertive
Foraging Ability: Excellent
Ideal Climate: Cold and damp
Notes & Tips: Polled. Long-haired. Efficient grazers.

critical

Canadienne

Origin: Quebec
Purpose: Dairy
Size (M/F, lbs.): 1,600/1,000
Color: Nearly black
Owner Skill: Intermediate

Temperament: Docile to assertive
Foraging Ability: Average
Ideal Climate: Cold and damp
Notes & Tips: Hardy and long-lived

study

Criollo (Mexican)

Origin: North Central Mexico

Purpose: Beef

Size (M/F, lbs.): 800–1,200/600–800

Color: Variable

Owner Skill: Intermediate to advanced

Temperament: Variable

Foraging Ability: Excellent

Ideal Climate: Hot and dry

Notes & Tips: Long-horned and athletic for rodeo stock

recovering

Devon (Beef Devon)

Origin: England

Purpose: Beef, oxen

Size (M/F, lbs.): 1,600–2,000/1,100

Color: Ruby red

Owner Skill: Novice

Temperament: Docile

Foraging Ability: Excellent

Ideal Climate: Temperate

Notes & Tips: Hardy and efficient for grass-based production

recovering

Dexter

Origin: Ireland

Purpose: Dairy, beef

Size (M/F, lbs.): Under 1,000/under 700

Color: Black, red, dun

Owner Skill: Novice to intermediate

Temperament: Docile

Foraging Ability: Average

Ideal Climate: Temperate

Notes & Tips: Dwarf dual-purpose cattle

critical

Dutch Belted

Origin: Netherlands

Purpose: Dairy

Size (M/F, lbs.): 1,600/1,000

Color: White belt on black or red

Owner Skill: Novice to intermediate

Temperament: Docile

Foraging Ability: Average

Ideal Climate: Temperate

Notes & Tips: Long lived. Excellent milk producer (20,000 lbs/year).

critical

Florida Cracker

Origin: Florida

Purpose: Beef

Size (M/F, lbs.): 800–1,200/600–800

Color: Variable

Owner Skill: Novice to intermediate

Temperament: Variable

Foraging Ability: Excellent

Ideal Climate: Hot and humid

Notes & Tips: Excellent production on poor range. Long-lived. Some are dwarf.

watch

Galloway

Origin: Scotland

Purpose: Beef

Size (M/F, lbs.): 1,800–2,000/1,200

Color: Black, red, dun, belted, white

Owner Skill: Novice to intermediate

Temperament: Active

Foraging Ability: Excellent

Ideal Climate: Cold

Notes & Tips: Shaggy beef cattle; forage well

watch

Guernsey

Origin: Isle of Guernsey

Purpose: Dairy

Size (M/F, lbs.): 2,000/1,400

Color: Gold and white

Owner Skill: Novice to intermediate

Temperament: Docile

Foraging Ability: Average

Ideal Climate: Temperate

Notes & Tips: High-protein milk

recovering

Highland

Origin: Scotland

Purpose: Beef

Size (M/F, lbs.): 1,500–2,000/900–1,300

Color: Gold, red, black, silver

Owner Skill: Novice to intermediate

Temperament: Variable

Foraging Ability: Excellent

Ideal Climate: Cold

Notes & Tips: Shaggy, long-horned cattle. Efficient. Rugged.

critical

Kerry

Origin: Ireland

Purpose: Dairy

Size (M/F, lbs.): 1,000/800

Color: Black

Owner Skill: Novice to intermediate

Temperament: Docile to assertive

Foraging Ability: Excellent

Ideal Climate: Cold

Notes & Tips: Small and rugged. Active.

critical

Milking Devon

Origin: England, then United States

Purpose: Dairy, beef, draft

Size (M/F, lbs.):
1,600–1,800/1,000–1,200

Color: Ruby red

Owner Skill: Intermediate

Temperament: Active, assertive

Foraging Ability: Excellent

Ideal Climate: Temperate

Notes & Tips: Active and athletic. High-protein milk. Good foragers.

critical

Milking Shorthorn – native

Origin: England
Purpose: Dairy, beef, draft
Size (M/F, lbs.):
1,800–2,000/1,100–1,400
Color: Red, red roan, white

Owner Skill: Novice
Temperament: Docile
Foraging Ability: Excellent
Ideal Climate: Temperate
Notes & Tips: Forage well; efficient

threatened

Pineywoods

Origin: Southern United States
Purpose: Beef
Size (M/F, lbs.): 800–1,200/600–800
Color: Variable
Owner Skill: Novice to intermediate
Temperament: Docile to active

Foraging Ability: Excellent
Ideal Climate: Hot; damp
Notes & Tips: Productive on poor
pasture. Long-lived. Some polled. Some
dwarf.

critical

Randall (Randall Lineback)

Origin: Vermont
Purpose: Beef, rose veal, oxen
Size (M/F, lbs.): 1,600/1,100
Color: Black or red with white lineback
Owner Skill: Intermediate to advanced

Temperament: Docile to assertive
Foraging Ability: Average
Ideal Climate: Cold
Notes & Tips: Originally triple purpose,
but few milked today

threatened

Red Poll

Origin: England
Purpose: Beef
Size (M/F, lbs.): 1,800/1,200
Color: Red
Owner Skill: Novice

Temperament: Docile
Foraging Ability: Excellent
Ideal Climate: Temperate
Notes & Tips: Polled. Efficient. Very
fertile. Long-lived.

critical

Texas Longhorn (CTLR)

Origin: Texas
Purpose: Beef
Size (M/F, lbs.): 800–1,200/600–800
Color: Variable
Owner Skill: Novice to intermediate
Temperament: Docile to active

Foraging Ability: Excellent
Ideal Climate: Hot and dry
Notes & Tips: Efficient range cattle,
valued for their longevity, fertility, and
adaptability

Horses

Akhal-Teke

Origin: Turkmenistan
Height: 15.2 hands (62")
Weight (M/F, lbs.): 900–1,000/900
Uses & Traits: Endurance and sport riding

Colors: Wide variation, with unique metallic sheen
Temperament: Active and spirited
Owner Skill: Intermediate to advanced
Notes & Tips: Spirited; bonds closely with rider

American Cream Draft Horse

Origin: Midwestern United States
Height: 15–16.3 hands (60–67")
Weight (M/F, lbs.): 1,800–2,000/ 1,600–1,800
Uses & Traits: Farm draft

Color: Cream with pink skin and amber eyes
Temperament: Docile
Owner Skill: Beginner to intermediate
Notes & Tips: America's only local draft horse breed

Canadian

Origin: Canada
Height: 14–16 hands (56–64")
Weight (M/F, lbs.): 900–1,000/ 900
Uses & Traits: Light driving and general riding

Colors: Black, brown, dark bay, rarely chestnut
Temperament: Willing worker
Owner Skill: Beginner to intermediate
Notes & Tips: A useful all-rounder with outstanding hardiness

Caspian

Origin: Iran
Height: 11.2–12.2 hands (46–50")
Weight (M/F, lbs.): 600/400
Uses & Traits: Refined child's riding pony
Colors: Bay, chestnut, black, gray, buckskin

Temperament: Docile
Owner Skill: Beginner
Notes & Tips: A small, refined, and very friendly pony

critical

Cleveland Bay

Origin: England
Height: 16–17 hands (64–68")
Weight (M/F, lbs.): 1,500/1,200
Uses & Traits: Sport riding and driving
Colors: Bay

Temperament: Docile and willing
Owner Skill: Beginner to intermediate
Notes & Tips: An old, useful, athletic breed

watch

Clydesdale

Origin: Scotland
Height: 16.2–18 hands (66–72")
Weight (M/F, lbs.): 1,800–2,000/ 1,600–1,800
Uses & Traits: Farm and city draft

Colors: Bay, black, chestnut, all with white markings
Temperament: Docile and willing
Owner Skill: Intermediate
Notes & Tips: High action and feathered feet

threatened

Colonial Spanish

Origin: United States
Height: 13–15.2 hands (52–62")
Weight (M/F, lbs.): 700–900/700–900
Uses & Traits: Riding, endurance
Colors: All colors

Temperament: Docile to alert
Owner Skill: Beginner
Notes & Tips: Many are gaited; most have willing attitudes

threatened

Dales Pony

Origin: England
Height: 14–14.2 hands (56–58")
Weight (M/F, lbs.): 800–1,000/ 800–900
Uses & Traits: Stout driving and riding pony

Colors: Black, brown, gray, roan
Temperament: Docile with spirit
Owner Skill: Beginner to intermediate
Notes & Tips: Stout conformation and distinctive feathered legs

threatened

Dartmoor

Origin: England
Height: 12 hands (48")
Weight (M/F, lbs.): 500/ 450–500
Uses & Traits: Child's riding pony
Colors: Brown, bay, black, rarely chestnut or gray

Temperament: Docile with spirit
Owner Skill: Beginner
Notes & Tips: Often used as a child's first pony

threatened

Exmoor

Origin: England
Height: 11.2 hands (46")
Weight (M/F, lbs.): 500/500
Uses & Traits: Primitive riding pony
Colors: Solid bay or brown

Temperament: Docile with spirit
Owner Skill: Beginner
Notes & Tips: Rugged and used to living outdoors

watch

Fell Pony

Origin: England
Height: 13–14 hands (52–56")
Weight (M/F, lbs.): 900/700
Uses & Traits: general riding, driving
Colors: Black, brown, bay, gray

Temperament: Docile with spirit
Owner Skill: Beginner to intermediate
Notes & Tips: Feathered feet; stout conformation

study

Galiceño

Origin: Mexico
Height: 12–13.2 hands (48–54")
Weight (M/F, lbs.): 600–800
Uses & Traits: Riding, ranch work
Colors: Solid colors, roan, and gray

Temperament: Docile and willing
Owner Skill: Beginner
Notes & Tips: A small but strong mount that can do the work of a horse

watch

Gotland

Origin: Gotland, Sweden
Height: 11.2–13 hands (46–52")
Weight (M/F, lbs.): 550
Uses & Traits: Riding pony

Colors: Black, bay, chestnut, gray, buckskin, palomino, dun
Temperament: Docile
Owner Skill: Beginner
Notes & Tips: A hardy riding and driving pony

critical

Hackney Horse

Origin: England
Height: 15–16 hands (60–64")
Weight (M/F, lbs.): 1,000–1,200
Uses & Traits: Driving and riding
Colors: Black, brown, bay, chestnut

Temperament: Spirited
Owner Skill: Beginner to intermediate
Notes & Tips: Flashy action; used for carriage driving

watch

Irish Draught

Origin: Ireland

Height: 15–17 hands (60–68")

Weight (M/F, lbs.): 1,150–1,600

Uses & Traits: Sport and event riding

Colors: Any solid color, usually dark

Temperament: Docile

Owner Skill: Beginner to intermediate

Notes & Tips: Used mostly in riding competition rather than driving

threatened

Lipizzan

Origin: Austria

Height: 14.2–15.2 hands (56–62")

Weight (M/F, lbs.): 1,100

Uses & Traits: Riding and driving

Colors: Gray, rarely black, or bay

Temperament: Docile and willing

Owner Skill: Beginner

Notes & Tips: Famous for high school dressage work

study

Morgan — traditional

Origin: New England, United States

Height: 14.1–15.2 hands (57–62")

Weight (M/F, lbs.): 800–1,000

Uses & Traits: Riding and driving

Colors: Bay, chestnut, black, rarely gray, roan, dun, buckskin, palomino, or spotted

Temperament: Variable; usually docile and willing

Owner Skill: Beginner to intermediate

Notes & Tips: The first American breed. Developed as an "all-rounder."

watch

Mountain Pleasure (Rocky Mountain)

Origin: Kentucky

Height: 14.2–15.2 hands (58–62")

Weight (M/F, lbs.): 850–1,000

Uses & Traits: Gaited riding horse

Colors: Any solid color; chocolate common

Temperament: Docile and willing

Owner Skill: Beginner

Notes & Tips: A distinctive gaited American breed with unusual colors

critical

Newfoundland Pony

Origin: Canada

Height: 11–14.2 hands (44–58")

Weight (M/F, lbs.): 400–750

Uses & Traits: Riding, driving, general work

Colors: Black, bay, chestnut, roan, gray

Temperament: Docile and willing

Owner Skill: Beginner

Notes & Tips: An old local breed now in low numbers

Shire

critical

Origin: England
Height: 17.2–19 hands (70–78")
Weight (M/F, lbs.): 1,800–2,100
Uses & Traits: Farm and city draft

Colors: Black, brown, gray, rarely bay or chestnut
Temperament: Docile
Owner Skill: Intermediate
Notes & Tips: Feathered feet and white markings distinguish this breed

Suffolk

critical

Origin: England
Height: 16–17 hands (64–68")
Weight (M/F, lbs.): 1,800
Uses & Traits: Farm draft
Colors: Always chestnut

Temperament: Docile and willing
Owner Skill: Novice to intermediate
Notes & Tips: A no-nonsense practical farm horse

Donkeys (Asses)

American Mammoth Jackstock

Origin: Upper South, United States
Height: Jacks, over 14.2 hands (58");
jennies, 14 hands (56")
Weight (M/F, lbs.): 900–1,200
Uses & Traits: Mule production, riding,
driving

Colors: Usually black or red, but all
colors allowed
Temperament: Docile to assertive
Owner Skill: Novice to intermediate;
advanced with jacks
Notes & Tips: The premier mule-
breeding jack in the world

Miniature Donkey

Origin: Italy
Height: Under 9 hands (36")
Weight (M/F, lbs.): 200–350
Uses & Traits: Riding; driving; pets

Colors: Gray dun usually, any and all
other colors
Temperament: Docile
Owner Skill: Beginner
Notes & Tips: Has turned the corner
from rarity

Poitou

Origin: France
Height: 14–15 hands (56–60")
Weight (M/F, lbs.): 750–950
Uses & Traits: Mule breeding
Colors: Black-brown with light points

Temperament: Docile
Owner Skill: Advanced
Notes & Tips: Young foals are delicate

Glossary

Breeding

Many words that breeders use have specific definitions that are used only for animal breeding. These terms can be confusing, especially when experienced folks use them at a rapid clip. Here is a list of some of the more common ones that are used fairly generally throughout several species and breeds.

BLOODLINE. A subfamily of a breed

BREED. A group of animals with obvious similarities that reproduces the same type of animal when bred to one another

CROSSBREEDING. The mating of animals of two different breeds

FLOCK. A group of animals of a single species. Usually used for sheep, chickens, ducks, geese, and turkeys.

HERD. A group of animals of a single species. Usually used for cattle, horses, asses, goats, swine, and rabbits

INBREEDING. The mating of animals that are closely related

LINEBREEDING. The mating of animals that are related, but not very closely

LIVESTOCK. Animals kept by people for use or pleasure

OUTBREEDING. The mating of animals that are not related

OUTCROSSING. The mating of animals that are not related, and that are from two different bloodlines within a single breed

STRAIN. A subfamily of a breed

TRIO. A group of poultry with one male and two females

VARIETY (OF POULTRY). Birds of a breed that vary from others by having a different comb type or color type

Cattle

BOVINE. Relating to cattle

BULL. An intact male bovine

BULL CALF. A male calf

CALF. A young bovine. This term is used from the time of birth up until about 6 to 10 months of age when the animal is weaned.

CATTLE. Domesticated mammals of the genus *Bos,* used for milk, meat, and draft power

COW. A female bovine that has had a calf

HEIFER. A female bovine that has not yet had a calf

HEIFER CALF. A female calf

OX. A castrated bull that has been trained to work and is at least 4 years of age

OXEN. The plural of ox

STEER. A castrated male bovine; or a future ox that is less than 4 years old

Chickens

BROILER. A meat chicken processed at the age of 7 to 12 weeks when it reaches 2½ to 3½ pounds live weight. Historically, birds ranging from 1 to 2½ pounds

CAPON. A male chicken that is castrated at 4 to 8 months old, weighing 5 to 9 pounds, and that produces more white meat and has higher fat content than other chickens

CHICK. A newly hatched or a very young chicken

CHICKEN. The common domestic fowl (*Gallus domesticus*), raised for eggs and meat

COCK. A male chicken at least 1 year of age or older

COCKEREL. A male chicken less than 1 year old

CORNISH GAME HENS. Cornish chickens of either sex, processed at a young age weighing about 1 to 2 pounds. Historically referred to as poussin or guinea hen.

FRYER. A meat chicken usually marketed at 12 to 20 weeks. Historically, birds ranging from 2½ to 3½ pounds

HEN. A female chicken at least 1 year of age

POULARDE. A fattened pullet primarily used as a roasting fowl

POUSSIN, SPRING CHICKEN. A young chicken, 3 to 4 weeks old, weighing about 1 pound, that is prepared as a single serving

PULLET. A female chicken less than 1 year of age. A pullet is, in industry, a young female that has yet to start laying eggs.

ROASTER. A chicken 6 to 12 months of age weighing 4 to 7 pounds

ROCK CORNISH GAME HEN. A cross between Cornish and Plymouth Rock chickens popularized in the 1950s in the United States. They are young birds, 4 to 5 weeks of age, of either sex and weigh approximately 2 pounds.

ROOSTER. A male chicken over 1 year of age

STEWING FOWL. A mature male or female chicken over 1 year of age

Donkeys (Asses)

ASS. A single-hoofed mammal of the genus *Equus*, related to horses, typically with a smaller build and longer ears. Ass is the correct term for a donkey, burro, or jackstock.

BURRO. The Spanish term for donkey (generally used west of the Mississippi)

EQUINE. Relating to horses, mules, zebras, and asses

HINNY. A hybrid animal produced when a female ass (jennet) is crossed with a male horse (stallion). Horse hinny is the proper term for a male hinny over 3 years of age; mare hinny is the proper term for a female hinny over 3 years of age.

JACK. A male ass. (Note: Young jacks are not referred to as colts.)

JACKSTOCK. The plural noun referring to American Mammoth jacks and jennets. These animals are properly termed asses, not referred to as donkeys, and never called burros.

JENNET, JENNY. A female ass. (Note: Young jennets or jennies are not referred to as fillies.)

JENNET JACK. A male ass used to produce donkeys

JOHN. An informal term for a male mule

MOLLY. An informal term for a female mule

MULE. A hybrid animal produced when a male ass (jack) is crossed with a female horse (mare). Mule colts (males) and mule fillies (females) are young mules under 3 years of age. Mare mules (females) and horse mules (males) are mules over 3 years of age. See also **hinny**

MULE JACK. A male ass used to produce mules

Ducks

DRAKE. An adult male duck

DUCK. Any wild or domesticated swimming bird of the family Anatidae, typically having a broad, flat bill, short legs, and webbed feet; a female of the duck family

DUCKLINGS. Baby ducks

OLD DRAKE. A male duck over 1 year of age

OLD DUCK. A female duck over 1 year of age

YOUNG DRAKE. A male duck under 1 year of age

YOUNG DUCK. A female duck under 1 year of age

Geese

GANDER. A male goose over 1 year of age. (Also referred to as an old gander.)

GEESE. Wild or domesticated water birds of the family Anatidae and the genera *Anser* and *Branta*, typically with a shorter neck than a swan and a shorter, more pointed bill than a duck

GOOSE. The singular of geese; a female goose

GOSLING. A young goose until the age when feathers have replaced all of its down

OLD GOOSE. A female goose over 1 year of age

YOUNG GANDER. A male goose under 1 year of age

YOUNG GOOSE. A female goose under 1 year of age

Goats

BUCK, BILLY. A male goat over 1 year of age

BUCKLING. A young male goat less than 1 year old

CAPRINE. Relating to goats

DOE, NANNY. A female goat over 1 year of age

DOELING. A young female goat less than a year old

GOAT. The domesticated form of *Capra hircus,* used for meat, milk, and fiber

KID. A baby goat of either sex

WETHER. A castrated male goat (or sheep)

Horses

COLT. A male horse under 3 years of age

EQUINE. Relating to horses, mules, zebras, and asses

FILLY. A female horse under 3 years of age

FOAL. A young equine under 1 year of age

GELDING. A castrated male horse; the act of castrating

HORSE. A domesticated, large, single-hoofed mammal (*Equus caballus*) with a short-haired coat, long mane, and long tail, used for riding, pulling, or carrying loads

MARE. A female horse after her 4th birthday

STALLION. An uncastrated (intact) male horse

Pigs

BARROW. A male swine that is castrated before sexual maturity

BOAR. An adult male swine

FEEDER PIG. A young pig, most often between 40 and 70 pounds, that is produced by one farmer and sold to another to grow out to market weight

GILT. A female swine that has not yet given birth

HOG. A mature swine with an adult weight above 150 pounds

LARD. The white solid or semisolid fat of a hog that is rendered and clarified for use in cooking

MARKET HOG. A hog that weighs from 220 to 260 pounds and is 5 to 7 months of age when sent to market

PIG. A young swine that is not sexually mature; or a mature swine with an adult weight under 150 pounds

PIGLET, BABY PIG. A young pig in its first 14 to 21 days of life and still nursing

PORCINE. Relating to swine

SHOAT. A young hog (not sexually mature) that has been weaned and is ready for market, weighing 150 to 260 pounds

SOW. An adult female swine

STAG. A male swine that is castrated after sexual maturity

SWINE. Even-toed ungulates of the family Suidae, including pigs, hogs, and boars, raised primarily for meat

Rabbits

BUCK. A male rabbit

DOE. A female rabbit

FRYER, YOUNG RABBIT. A rabbit that is 2 months old and weighs 3¾ to 4½ pounds

KIT. A baby rabbit

RABBIT. A mammal of the family Leporidae or the domesticated Old World species *Oryctolagus cuniculus*

RABBITRY. A facility housing a herd of rabbits in separate cages

STEWER, MATURE RABBIT. A rabbit 3 months of age or older averaging 6 pounds or more

WARREN. A large cage or enclosure keeping a herd of rabbits as a group

Sheep

EWE. A female sheep at least 1 year of age

EWE LAMB. A female sheep under 1 year of age

LAMB. A young sheep. When referring to meat, lamb is meat from a sheep that is 12 to 14 months old or younger.

LAMBKIN, LAMBLING. A newborn lamb

OVINE. Relating to sheep

RAM. An intact male sheep at least 1 year of age

RAM LAMB. A male sheep under 1 year of age

SHEEP. A domesticated member of the species *Ovis aries*, used for wool, meat, and milk; a mature ovine at least 1 year of age

WETHER. A castrated male sheep (or goat)

Turkeys

HEN. A female turkey

OLD HEN. A female turkey over 1 year of age

OLD TOM. A male turkey over 1 year of age

POULT. A young domestic (not wild) turkey

TOM. A male turkey

TURKEY. A large North American bird (*Meleagris gallopavo*) that is widely domesticated for food production and comes in many varieties

YOUNG HEN. A female turkey under 1 year of age

YOUNG TOM. A male turkey under 1 year of age

Resources

These resources are by no means comprehensive, but they will give you a start when searching for general information about succeeding with livestock and poultry on your farm or ranch.

Beginning Farmers' Resources and Classes

MORE FARMERS ON THE LAND
Land Stewardship Project
http://landstewardshipproject.org/morefarmers

NEW ENGLAND SMALL FARM INSTITUTE
www.smallfarm.org

NEW FARMER HUB
Northeast Beginning Farmers Project
http://nebeginningfarmers.org/farmers

Online Publications for Sustainable Practices

KERR CENTER PUBLICATIONS
Kerr Center
http://www.kerrcenter.com/resources/kerr-center-publications.htm
Sustainable forage practices and more.

MASTER PUBLICATION LIST
National Center for Appropriate Technology (ATTRA)
https://attra.ncat.org/publication.html
Production, starting a farm, links to other resources.

SUSTAINABLE AGRICULTURE RESEARCH & EDUCATION
www.sare.org
Their learning center has bulletins on production, starting a farm, and marketing.

Books on Conservation Breeding Practices

Commission on Genetic Resources for Food and Agriculture. *Breeding Strategies for Sustainable Management of Animal Genetic Resources.* Food and Agriculture Organization of the United Nations, Animal Production and Health Guidelines No. 3, 2010.

Sponenberg, D. Phillip, and Carolyn J. Christman. *A Conservation Breeding Handbook.* American Livestock Breeds Conservancy, 1995.

Sponenberg, D. Phillip, and Donald E. Bixby. *Managing Breeds for a Secure Future.* American Livestock Breeds Conservancy, 2007.

Index

To find a particular breed, look under its species: for example, look under *Horses* for *Shire*. A **boldfaced** and *italicized* page number indicates that a breed summary appears on that page.

Other Storey Titles You Will Enjoy

The Backyard Homestead *edited by Carleen Madigan*
A complete guide to growing and raising the most local food
available anywhere — from one's own backyard.
368 pages. Paper. ISBN 978-1-60342-138-6.

The Backyard Homestead Guide to Raising Farm Animals
edited by Gail Damerow
Expert advice on raising healthy, happy, productive farm animals.
360 pages. Paper. ISBN 978-1-60342-969-6.

Oxen: A Teamster's Guide *by Drew Conroy*
The definitive guide to selecting, training, and caring for the mighty ox.
304 pages. Paper. ISBN 978-1-58017-692-7.

Reclaiming Our Food *by Tanya Denckla Cobb*
Stories of more than 50 groups across the United States that are finding
innovative ways to provide local food to their communities.
320 pages. Paper. ISBN 978-1-60342-799-9.

Storey's Guide to Raising Series
Everything you need to know to keep your livestock and your profits healthy.
Please visit *www.storey.com* for a full list of titles.

Storey's Illustrated Breed Guide to Sheep, Goats, Cattle, and Pigs
by Carol Ekarius
A comprehensive, colorful, and captivating in-depth guide to North America's
common and heritage breeds.
320 pages. Paper. ISBN 978-1-60342-036-5.
Hardcover with jacket. ISBN 978-1-60342-037-2.

Storey's Illustrated Guide to Poultry Breeds *by Carol Ekarius*
A definitive presentation of more than 120 barnyard fowl, complete with
full-color photographs and detailed descriptions.
288 pages. Paper. ISBN 978-1-58017-667-5.

These and other books from Storey Publishing are available
wherever quality books are sold or by calling 1-800-441-5700.
Visit us at *www.storey.com* or sign up for our newsletter
at *www.storey.com/signup.*